Android企业级项目实战教程

黑马程序员 / 编著

清华大学出版社
北京

内 容 简 介

本书适合有一定 Android 基础的读者阅读，书中涵盖多个技术热点，其中包括下拉刷新、HelloCharts 图表库、BoomMenu 圆形菜单、BubbleViews 心形泡泡库、第三方视频播放等。本书在结构设计上采用由小项目逐渐深入的形式，然后引出一个黑马头条项目，讲解企业项目的开发流程。本书共 13 章，第 1～4 章每章分别讲解一个小项目，第 5～13 章每章分别讲解黑马头条项目的一个模块，包括从项目分析、效果展示到项目开发、打包发布的全过程。

本书既可作为高等院校本、专科计算机相关专业的教材，也可作为社会培训教材，是一本适合广大编程爱好者参考和学习的书籍。

本书封面贴有清华大学出版社防伪标签，无标签者不得销售。
版权所有，侵权必究。举报：010-62782989，beiqinquan@tup.tsinghua.edu.cn。

图书在版编目(CIP)数据

Android 企业级项目实战教程/黑马程序员编著. —北京：清华大学出版社，2018(2023.8重印)
ISBN 978-7-302-49120-0

Ⅰ. ①A… Ⅱ. ①黑… Ⅲ. ①移动终端－应用程序－程序设计－教材 Ⅳ. ①TP929.53

中国版本图书馆 CIP 数据核字(2017)第 315395 号

责任编辑：袁勤勇
封面设计：韩 冬
责任校对：焦丽丽
责任印制：沈 露

出版发行：清华大学出版社
　　　　网　　址：http://www.tup.com.cn，http://www.wqbook.com
　　　　地　　址：北京清华大学学研大厦 A 座　　邮　　编：100084
　　　　社 总 机：010-83470000　　邮　　购：010-62786544
　　　　投稿与读者服务：010-62776969，c-service@tup.tsinghua.edu.cn
　　　　质量反馈：010-62772015，zhiliang@tup.tsinghua.edu.cn
　　　　课件下载：http://www.tup.com.cn，010-83470236
印 装 者：三河市铭诚印务有限公司
经　　销：全国新华书店
开　　本：185mm×260mm　　印　　张：21.75　　字　　数：525 千字
版　　次：2018 年 2 月第 1 版　　印　　次：2023 年 8 月第 7 次印刷
定　　价：59.00 元

产品编号：077936-03

序 言

江苏传智播客教育科技股份有限公司(简称"传智播客")是一家致力于培养高素质软件开发人才的科技公司。经过多年探索,传智播客的战略逐步完善,从 IT 教育培训发展到高等教育,从根本上解决以"人"为单位的系统教育培训问题,实现新的系统教育形态,构建出前后衔接、相互呼应的分层次教育培训模式。

一、"黑马程序员"——高端 IT 教育品牌

"黑马程序员"的学员多为大学毕业后,想从事 IT 行业,但各方面条件还不成熟的年轻人。"黑马程序员"的学员筛选制度非常严格,包括了严格的技术测试、自学能力测试,以及性格测试、压力测试、品德测试等。百里挑一的残酷筛选制度确保学员质量,并降低企业的用人风险。

自"黑马程序员"成立以来,教学研发团队一直致力于打造精品课程资源,不断在产、学、研 3 个层面创新自己的执教理念与教学方针,并集中"黑马程序员"的优势力量,有针对性地出版了计算机系列教材 90 多种,制作教学视频数十套,发表各类技术文章数百篇。

"黑马程序员"不仅斥资研发 IT 系列教材,还为高校师生提供以下配套学习资源与服务。

1. 为大学生提供的配套服务

(1) 请同学们登录 http://yx.ityxb.com,进入"高校学习平台",免费获取海量学习资源。平台可以帮助高校学生解决各类学习问题。

(2) 针对高校学生在学习过程中存在的压力大等问题,我们还面向大学生量身打造了 IT 技术女神——"播妞学姐",可提供教材配套源码、习题答案及更多学习资源。同学们快来关注"播妞学姐"的微信公众号 boniu1024。

"播妞学姐"微信公众号

2. 为教师提供的配套服务

针对高校教学,"黑马程序员"为 IT 系列教材精心设计了"教案+授课资源+考试系统+题库+教学辅助案例"的系列教学资源。高校老师请登录 http://yx.ityxb.com,进入"高校教辅平台",也可关注"码大牛"老师微信/QQ:2011168841,获取配套资源,还可以扫

描下方二维码,关注专为 IT 教师打造的师资服务平台——"教学好助手",获取最新的教学辅助资源。

"教学好助手"微信公众号

二、"传智专修学院"——高等教育机构

传智专修学院是一所由江苏省宿迁市教育局批准、江苏传智播客教育科技股份有限公司投资创办的四年制应用型院校。学校致力于为互联网、智能制造等新兴行业培养高精尖科技人才,聚焦人工智能、大数据、机器人、物联网等前沿技术,开设软件工程专业,招收的学生入校后将接受系统化培养,毕业时学生的专业水平和技术能力可满足大型互联网企业的用人要求。

传智专修学院借鉴卡内基·梅隆大学、斯坦福大学等世界著名大学的办学模式,采用"申请入学,自主选拔"的招生方式,通过深入调研企业需求,以校企合作、专业共建等方式构建专业的课程体系。传智专修学院拥有顶级的教研团队、完善的班级管理体系、匠人精神的现代学徒制和敢为人先的质保服务。

传智专修学院突出的办学特色如下。

(1)立足"高精尖"人才培养。传智专修学院以国家重大战略和国际科学技术前沿为导向,致力于为社会培养具有创新精神和实践能力的应用型人才。

(2)项目式教学,培养学生自主学习能力。传智专修学院打破传统高校理论式教学模式,将项目实战式教学模式融入课堂,通过分组实战,模拟企业项目开发过程,让学生拥有真实的工作能力,并持续培养学生的自主学习能力。

(3)创新模式,就业无忧。学校为学生提供"一年工作式学习",学生能够进入企业边工作边学习。与此同时,我们还提供专业老师指导学生参加企业面试,并且开设了技术服务窗口给学生解答工作中遇到的各种问题,帮助学生顺利就业。

如果想了解传智专修学院更多的精彩内容,请关注微信公众号"传智专修学院"。

传智专修学院

传智播客

2020 年 2 月

前 言

为什么要学习 Android

Android 是 Google 公司开发的基于 Linux 的开源操作系统，主要应用于智能手机、平板电脑等移动设备，还可用于穿戴设备、智能家居等领域。经过短短几年的发展，Android 系统在全球得到了大规模推广。据不完全统计，Android 系统已经用于全球 80％以上的智能手机，中国市场的占有率更是高达 90％以上，因此越来越多的人开始学习 Android 技术，以适应市场需并寻求更广阔的发展空间。

本书在修订过程中，结合党的二十大精神"进教材、进课堂、进头脑"的要求，将知识教育与思想政治教育相结合，通过案例加深学生对知识的认识与理解，注重培养学生的创新精神、实践能力和社会责任感。案例设计从现实需求出发，激发学生的学习兴趣，充分发挥学生的主动性和积极性，增强学习信心和学习欲望，培养学生分析问题和解决问题的能力。在知识和案例的讲解中融入了素质教育的相关内容，引导学生树立正确的世界观、人生观和价值观，进一步提升学生的职业素养，落实德才兼备的高素质卓越工程师和高技能人才的培养要求。此外，作者依据书中的内容提供了线上学习资源，体现现代信息技术与教育教学的深度融合，进一步推动教育数字化发展。

如何使用本书

本书适合有一定 Android 基础的读者阅读，书中涵盖多个技术热点，其中包括下拉刷新、HelloCharts 图表库、BoomMenu 圆形菜单、BubbleViews 心形泡泡库、第三方视频播放等。若读者没有 Android 基础，建议读者先学习《Android 移动开发基础案例教程》，然后再学习本书。

本书在结构设计上采用由小项目逐渐深入的形式，然后引出一个黑马头条项目，讲解企业项目的开发流程。本书共 13 章，第 1~4 章分别讲解一个小项目，第 5~13 章讲解黑马头条项目。具体如下。

- 第 1 章主要讲解弹幕项目的实现过程，其中包括弹幕下方的视频播放、弹幕文本的发送等。
- 第 2 章主要讲解 VR 项目的实现过程，其中包括全景图片与全景视频介绍、VR 全景图片的预览、VR 全景视频的播放等。

- 第 3 章主要讲解 HelloCharts 图表库的使用，其中以饼状图、柱状图、线形图为例进行演示。
- 第 4 章主要讲解 3D 相册的实现过程，其中包括图片展示与滑动效果等。
- 第 5~13 章分别讲解黑马头条项目的各个模块，其中包括项目综述、欢迎模块、首页模块、统计模块、视频模块、"我"模块、设置模块和项目上线。

读者在阅读本书的过程中难免会遇到一些问题，如果是对某个知识点不熟悉，则可以先行查阅，然后再进行项目开发。黑马头条项目是一个完整的项目，建议读者先厘清思路，多思考、多分析、多实践，逐步完成项目的开发。

如何领取课程

1. 扫描封面底部或右侧二维码，注册博学谷账号；
2. 点击底部导航栏"个人"，选择"我的优惠券"；
3. 输入封底刮刮卡中的优惠码，即可兑换相应的课程。

扫码领取课程

致谢

本书的编写和整理工作由传智播客教育科技股份有限公司完成，主要参与人员有张泽华、李印东、邱本超、殷凯等，全体参编人员在将近一年的编写过程中付出了很多辛勤的汗水，在此一并表示衷心的感谢。

意见反馈

尽管我们尽了最大的努力，但本书中难免存在不妥之处，欢迎各界专家和读者朋友们来信来函给予宝贵意见，我们将不胜感激。读者在阅读本书时，如发现任何问题或有不认同之处，可以通过发送电子邮件与我们取得联系。

请发送电子邮件至 itcast_book@vip.sina.com。

黑马程序员
2023 年 7 月于北京

目　录

第1章　弹幕 ·· 1

　1.1　弹幕程序 ··· 1
　　　任务综述 ·· 1
　　　【任务1-1】弹幕界面 ··· 1
　　　【任务1-2】弹幕界面逻辑代码 ·· 3
　1.2　本章小结 ··· 7

第2章　VR ·· 8

　2.1　全景图片与全景视频介绍 ··· 8
　2.2　VR主界面 ·· 9
　　　任务综述 ·· 9
　　　【任务2-1】VR主界面 ··· 9
　　　【任务2-2】VR主界面逻辑
　　　　　　　　代码 ·· 10
　2.3　VR全景图片 ·· 11
　　　任务综述 ··· 11
　　　【任务2-3】VR全景图片界面 ·· 11
　　　【任务2-4】VR全景图片界面
　　　　　　　　逻辑代码 ··· 13
　2.4　VR全景视频 ·· 14
　　　任务综述 ··· 14
　　　【任务2-5】VR全景视频界面 ·· 15
　　　【任务2-6】VR全景视频界面
　　　　　　　　逻辑代码 ··· 16
　2.5　本章小结 ·· 19

第3章　图表库 ·· 20

　3.1　线形图 ··· 20
　　　任务综述 ··· 20
　　　【任务3-1】线形图界面 ··· 20

		【任务 3-2】 创建 ViewPagerAdapter ···	22
		【任务 3-3】 线形图界面逻辑代码 ···	23
	3.2	饼状图 ··	26
		任务综述 ··	26
		【任务 3-4】 饼状图界面 ···	26
		【任务 3-5】 饼状图界面逻辑代码 ···	27
	3.3	柱状图 ··	30
		任务综述 ··	30
		【任务 3-6】 柱状图界面 ···	30
		【任务 3-7】 柱状图界面逻辑代码 ···	31
	3.4	本章小结 ··	33

第 4 章 3D 相册 ··· 34

	4.1	相册 ··	34
		任务综述 ··	34
		【任务 4-1】 "相册"界面 ··	34
		【任务 4-2】 "相册"界面 Item ···	36
		【任务 4-3】 创建 AlbumBean ··	38
		【任务 4-4】 "相册"界面 Adapter ··	38
		【任务 4-5】 "相册"界面逻辑代码 ··	39
	4.2	本章小结 ··	43

第 5 章 项目综述 ··· 44

	5.1	项目分析 ··	44
		5.1.1 项目名称 ···	44
		5.1.2 项目概述 ···	44
		5.1.3 开发环境 ···	44
		5.1.4 模块说明 ···	45
	5.2	效果展示 ··	45
		5.2.1 欢迎界面与主界面 ···	45
		5.2.2 "新闻详情"界面与"Python 学科"界面 ···	45
		5.2.3 "统计详情"界面 ···	46
		5.2.4 "视频详情"界面 ···	48
		5.2.5 "我"界面 ···	48
		5.2.6 "个人资料"界面 ···	51
	5.3	本章小结 ··	53

第 6 章 欢迎模块 ··· 54

	6.1	欢迎界面 ··	54

　　　　任务综述 ··· 54
　　　　【任务 6-1】 欢迎界面 ·· 54
　　　　【任务 6-2】 欢迎界面逻辑代码 ·· 55
　6.2 导航栏 ··· 56
　　　　任务综述 ··· 56
　　　　【任务 6-3】 标题栏 ·· 56
　　　　【任务 6-4】 底部导航栏 ··· 57
　　　　【任务 6-5】 底部导航栏逻辑代码 ·· 60
　6.3 本章小结 ·· 63

第 7 章 首页模块 ·· 64

　7.1 搭建服务器 ··· 64
　　　　任务综述 ··· 64
　　　　【任务 7-1】 首页广告栏数据 ·· 64
　　　　【任务 7-2】 首页新闻列表数据 ·· 66
　7.2 工具类 ··· 67
　　　　任务综述 ··· 67
　　　　【任务 7-3】 创建 Constant 类 ·· 67
　　　　【任务 7-4】 创建 JsonParse 类 ··· 68
　　　　【任务 7-5】 创建 UtilsHelper 类 ··· 68
　7.3 首页 ··· 69
　　　　任务综述 ··· 69
　　　　【任务 7-6】 水平滑动广告栏界面 ··· 69
　　　　【任务 7-7】 首页界面 ·· 72
　　　　【任务 7-8】 自定义控件 WrapRecyclerView ······························ 76
　　　　【任务 7-9】 首页界面 Item ··· 80
　　　　【任务 7-10】 创建 NewsBean ··· 83
　　　　【任务 7-11】 创建 AdBannerFragment ·································· 85
　　　　【任务 7-12】 创建 AdBannerAdapter ···································· 86
　　　　【任务 7-13】 首页界面 Adapter ·· 88
　　　　【任务 7-14】 首页界面逻辑代码 ·· 90
　7.4 新闻详情 ·· 96
　　　　任务综述 ··· 96
　　　　【任务 7-15】 "新闻详情"界面 ·· 97
　　　　【任务 7-16】 "新闻详情"界面逻辑代码 ··································· 99
　7.5 Python 学科 ··· 106
　　　　任务综述 ··· 106
　　　　【任务 7-17】 "Python 学科"界面 ·· 106
　　　　【任务 7-18】 "Python 学科"界面 Item ·································· 107

【任务 7-19】　创建 PythonBean ……………………………………………………… 108
　　　【任务 7-20】　"Python 学科"界面 Adapter ……………………………………… 109
　　　【任务 7-21】　"Python 学科"界面逻辑代码 ……………………………………… 110
　7.6　本章小结 ……………………………………………………………………………… 113

第 8 章　统计模块 …………………………………………………………………………… 115

　8.1　统计 …………………………………………………………………………………… 115
　　　任务综述 ……………………………………………………………………………… 115
　　　【任务 8-1】　"统计"界面 …………………………………………………………… 115
　　　【任务 8-2】　"统计"界面逻辑代码 ………………………………………………… 117
　8.2　统计详情 ……………………………………………………………………………… 120
　　　任务综述 ……………………………………………………………………………… 120
　　　【任务 8-3】　"Android 统计"详情界面 …………………………………………… 120
　　　【任务 8-4】　"Android 统计"详情界面逻辑代码 ………………………………… 122
　　　【任务 8-5】　"Java 统计"详情界面 ………………………………………………… 124
　　　【任务 8-6】　"Java 统计"详情界面逻辑代码 ……………………………………… 126
　8.3　本章小结 ……………………………………………………………………………… 130

第 9 章　视频模块 …………………………………………………………………………… 131

　9.1　视频列表 ……………………………………………………………………………… 131
　　　任务综述 ……………………………………………………………………………… 131
　　　【任务 9-1】　"视频列表"界面 ……………………………………………………… 131
　　　【任务 9-2】　"视频列表"界面 Item ………………………………………………… 132
　　　【任务 9-3】　创建 VideoBean ……………………………………………………… 133
　　　【任务 9-4】　"视频列表"界面 Adapter …………………………………………… 135
　　　【任务 9-5】　"视频列表"界面数据 ………………………………………………… 136
　　　【任务 9-6】　"视频列表"界面逻辑代码 …………………………………………… 138
　9.2　视频详情 ……………………………………………………………………………… 140
　　　任务综述 ……………………………………………………………………………… 140
　　　【任务 9-7】　"视频详情"界面 ……………………………………………………… 141
　　　【任务 9-8】　"视频目录"列表 Item ………………………………………………… 148
　　　【任务 9-9】　画面尺寸菜单 ………………………………………………………… 149
　　　【任务 9-10】　"视频目录"列表 Adapter …………………………………………… 151
　　　【任务 9-11】　创建 TopLineApplication …………………………………………… 154
　　　【任务 9-12】　创建 VideoDetailPagerAdapter ……………………………………… 155
　　　【任务 9-13】　创建 ParamsUtils ……………………………………………………… 156
　　　【任务 9-14】　视频播放进度条 ……………………………………………………… 157
　　　【任务 9-15】　画面尺寸菜单逻辑代码 ……………………………………………… 159
　　　【任务 9-16】　视频清晰度菜单逻辑代码 …………………………………………… 160

【任务 9-17】 "视频详情"界面逻辑代码 ································ 162

9.3 本章小结 ································ 164

第 10 章 "我"模块（一） ································ 166

10.1 创建数据库 ································ 166
  任务综述 ································ 166
  【任务 10-1】 创建 SQLite 数据库 ································ 166
  【任务 10-2】 创建 DBUtils 类 ································ 167
  【任务 10-3】 创建 UserBean ································ 168

10.2 "我" ································ 169
  任务综述 ································ 169
  【任务 10-4】 "我"界面 ································ 169
  【任务 10-5】 广播接收者 ································ 180
  【任务 10-6】 "我"界面逻辑代码 ································ 181

10.3 注册 ································ 185
  任务综述 ································ 185
  【任务 10-7】 "注册"界面 ································ 186
  【任务 10-8】 MD5 加密算法 ································ 192
  【任务 10-9】 "注册"界面逻辑代码 ································ 193

10.4 登录 ································ 196
  任务综述 ································ 196
  【任务 10-10】 "登录"界面 ································ 196
  【任务 10-11】 "登录"界面逻辑代码 ································ 199

10.5 个人资料 ································ 203
  任务综述 ································ 203
  【任务 10-12】 "个人资料"界面 ································ 203
  【任务 10-13】 "个人资料"界面逻辑代码 ································ 207

10.6 个人资料修改 ································ 214
  任务综述 ································ 214
  【任务 10-14】 个人资料修改界面 ································ 215
  【任务 10-15】 个人资料修改界面逻辑代码 ································ 216

10.7 本章小结 ································ 221

第 11 章 "我"模块（二） ································ 222

11.1 日历 ································ 222
  任务综述 ································ 222
  【任务 11-1】 "日历"界面 ································ 222
  【任务 11-2】 "日历"界面逻辑代码 ································ 224

11.2 星座 ································ 226

    任务综述 ……………………………………………………………………… 226
    【任务 11-3】 "星座"界面 …………………………………………………… 226
    【任务 11-4】 创建 ConstellationBean …………………………………………… 234
    【任务 11-5】 "星座"界面数据 …………………………………………………… 236
    【任务 11-6】 "星座"界面逻辑代码 ……………………………………………… 240
  11.3 星座选择 ……………………………………………………………………… 244
    任务综述 ……………………………………………………………………… 244
    【任务 11-7】 "星座选择"界面 …………………………………………………… 244
    【任务 11-8】 "星座选择"界面 Item ……………………………………………… 246
    【任务 11-9】 "星座选择"界面 Adapter …………………………………………… 247
    【任务 11-10】 "星座选择"界面数据 ……………………………………………… 249
    【任务 11-11】 "星座选择"界面逻辑代码 ………………………………………… 250
  11.4 涂鸦 …………………………………………………………………………… 252
    任务综述 ……………………………………………………………………… 252
    【任务 11-12】 "涂鸦"界面 ……………………………………………………… 252
    【任务 11-13】 涂鸦颜色选择界面 ………………………………………………… 264
    【任务 11-14】 创建 ColorsBean …………………………………………………… 271
    【任务 11-15】 创建 BigSizeBean ………………………………………………… 272
    【任务 11-16】 "涂鸦"界面逻辑代码 ……………………………………………… 273
  11.5 地图 …………………………………………………………………………… 280
    任务综述 ……………………………………………………………………… 280
    【任务 11-17】 "地图"界面 ……………………………………………………… 280
    【任务 11-18】 "地图"界面逻辑代码 ……………………………………………… 281
  11.6 本章小结 ……………………………………………………………………… 286

第 12 章 设置模块 ……………………………………………………………… 287

  12.1 收藏 …………………………………………………………………………… 287
    任务综述 ……………………………………………………………………… 287
    【任务 12-1】 "收藏"界面 ……………………………………………………… 287
    【任务 12-2】 "收藏"界面 Item ………………………………………………… 289
    【任务 12-3】 "收藏"界面 Adapter ……………………………………………… 294
    【任务 12-4】 收藏新闻信息表 …………………………………………………… 297
    【任务 12-5】 "收藏"界面逻辑代码 ……………………………………………… 300
  12.2 设置 …………………………………………………………………………… 302
    任务综述 ……………………………………………………………………… 302
    【任务 12-6】 "设置"界面 ……………………………………………………… 303
    【任务 12-7】 "设置"界面逻辑代码 ……………………………………………… 305
  12.3 修改密码 ……………………………………………………………………… 308
    任务综述 ……………………………………………………………………… 308

　　　　【任务12-8】"修改密码"界面 ……………………………………………………… 308
　　　　【任务12-9】"修改密码"界面逻辑代码 ………………………………………… 310
　12.4　设置密保 …………………………………………………………………………………… 313
　　　　任务综述 ………………………………………………………………………………… 313
　　　　【任务12-10】"设置密保"界面 ……………………………………………………… 314
　　　　【任务12-11】"设置密保"界面逻辑代码 …………………………………………… 316
　12.5　本章小结 …………………………………………………………………………………… 320

第13章　项目上线 …………………………………………………………………………………… 321

　13.1　代码混淆 …………………………………………………………………………………… 321
　　　　13.1.1　修改 build.gradle 文件 ………………………………………………………… 321
　　　　13.1.2　编写 proguard-rules.pro 文件 ………………………………………………… 321
　13.2　项目打包 …………………………………………………………………………………… 323
　13.3　项目加固 …………………………………………………………………………………… 326
　13.4　项目发布 …………………………………………………………………………………… 330
　13.5　本章小结 …………………………………………………………………………………… 332

第 1 章 弹 幕

学习目标
- 掌握弹幕界面的开发,能够独立实现弹幕程序

在程序开发中,弹幕用得最多的地方要数直播平台了,例如常见的斗鱼。在观看视频直播时,通常在屏幕底部会有一个输入框,专门用于输入弹幕。弹幕其实很像人们发送的即时消息,只不过是显示在直播界面上。本章让我们一起实现自己的弹幕程序吧!

1.1 弹幕程序

思政材料 1

任务综述

弹幕界面主要显示视频播放、弹幕文本信息、弹幕输入框以及弹幕的"发送"按钮,当点击弹幕界面上的任意地方时,界面底部会弹出一个输入框和一个"发送"按钮,此时可以输入文字并发送。当再次点击弹幕界面上的任意地方时,界面底部的输入框和"发送"按钮便会消失。为了与他人发送的弹幕有所区分,本机发送的弹幕会添加一个蓝色的边框。

【任务 1-1】 弹幕界面

【任务分析】

弹幕界面主要显示视频播放、弹幕文本信息、弹幕输入框以及弹幕的"发送"按钮,界面效果如图 1-1 所示。

【任务实施】

(1) 创建项目。首先创建一个项目,将其命名为 Barrage,指定包名为 com.itheima.barrage。将项目的 icon 图标 barrage_icon.png 导入 mipmap-hdpi 文件夹。

(2) 导入视频文件。在程序的 res 目录中创建一个 raw 目录,将视频文件 sun.mp4 导入该目录。

(3) 添加 DanmakuFlameMaster 库。由于弹幕界面中用到了 DanmakuFlameMaster 库中的 DanmakuView 控件,因此需要将 DanmakuFlameMaster 库添加到本项目中。在 Android Studio 中,选中项目,右键选择 Open Module Settings 选项后选择 Dependencies 选项卡,单击右上角的绿色加号,选择 Library dependency 选项,把 com.github.ctiao: DanmakuFlameMaster:0.5.3 库加入项目。

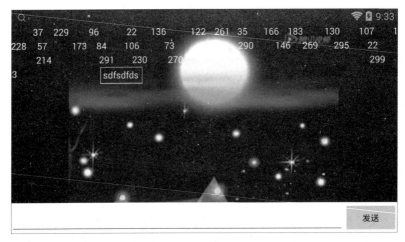

图 1-1 弹幕界面

注意：如果在 Library dependency 选项中找不到该库，则可以直接在该项目的 build.gradle 文件中添加如下代码。

```
compile 'com.github.ctiao:DanmakuFlameMaster:0.5.3'
```

（4）放置界面控件。在 activity_main.xml 文件中放置一个 VideoView 控件用于播放视频，一个 DanmakuView 控件用于显示弹幕，一个 EditText 控件用于输入弹幕文本，一个 Button 控件用于显示发送弹幕的按钮，具体代码如文件 1-1 所示。

【文件 1-1】 activity_main.xml

```
1   <?xml version="1.0" encoding="utf-8"?>
2   <RelativeLayout xmlns:android="http://schemas.android.com/apk/res/android"
3       android:id="@+id/activity_main"
4       android:layout_width="match_parent"
5       android:layout_height="match_parent"
6       android:background="#000">
7       <VideoView
8           android:id="@+id/videoview"
9           android:layout_width="match_parent"
10          android:layout_height="wrap_content"
11          android:layout_centerInParent="true" />
12      <master.flame.danmaku.ui.widget.DanmakuView
13          android:id="@+id/danmaku"
14          android:layout_width="match_parent"
15          android:layout_height="match_parent" />
16      <LinearLayout
17          android:id="@+id/ly_send"
18          android:layout_width="match_parent"
19          android:layout_height="50dp"
20          android:layout_alignParentBottom="true"
21          android:background="#fff"
22          android:visibility="gone">
```

```
23          <EditText
24              android:id="@+id/et_text"
25              android:layout_width="0dp"
26              android:layout_height="match_parent"
27              android:layout_weight="1" />
28          <Button
29              android:id="@+id/btn_send"
30              android:layout_width="wrap_content"
31              android:layout_height="match_parent"
32              android:text="发送" />
33      </LinearLayout>
34  </RelativeLayout>
```

【任务 1-2】 弹幕界面逻辑代码

【任务分析】

当点击弹幕界面时,界面底部会弹出一个输入框和一个"发送"按钮,当输入文字并点击"发送"按钮后,弹幕会出现一个蓝色框,框中显示刚刚发送的文字;当再次点击弹幕界面时,界面底部的输入框和"发送"按钮消失。

【任务实施】

(1) 获取界面控件。在 MainActivity 中创建一个 initView() 方法,用于获取弹幕界面上所要用到的控件。

(2) 播放视频。在 MainActivity 中创建一个 playVideo() 方法,在该方法中获取 res/raw 文件夹中的 sun.mp4 视频文件的 uri,并调用 setVideoURI() 方法设置 uri,然后调用 start() 方法播放视频。

(3) 初始化弹幕。在 MainActivity 中创建一个 initDanmaku() 方法,用于初始化弹幕并调用弹幕的随机生成与添加方法。

(4) 随机生成与添加弹幕。在 MainActivity 中创建 generateDanmakus() 与 addDanmaku() 方法,分别用于随机生成弹幕文字与添加弹幕到屏幕上,具体代码如文件 1-2 所示。

【文件 1-2】 MainActivity.java

```
1   package com.itheima.barrage;
2   public class MainActivity extends AppCompatActivity {
3       private boolean showDanmaku;
4       private DanmakuView danmakuView;
5       private DanmakuContext danmakuContext;
6       private Button sendButton;
7       private LinearLayout sendLayout;
8       private EditText editText;
9       private VideoView videoView;
10      @Override
11      protected void onCreate(Bundle savedInstanceState) {
12          super.onCreate(savedInstanceState);
```

```
13          setContentView(R.layout.activity_main);
14          initView();
15          playVideo();
16          initDanmaku();
17      }
18      /**
19       * 初始化界面控件
20       */
21      private void initView() {
22          videoView= (VideoView) findViewById(R.id.videoview);
23          sendLayout= (LinearLayout) findViewById(R.id.ly_send);
24          sendButton= (Button) findViewById(R.id.btn_send);
25          editText= (EditText) findViewById(R.id.et_text);
26          danmakuView= (DanmakuView) findViewById(R.id.danmaku);
27      }
28      /**
29       * 播放视频
30       */
31      private void playVideo() {
32          String uri="android.resource://"+getPackageName()+"/"+R.raw.sun;
33          if(uri !=null) {
34              videoView.setVideoURI(Uri.parse(uri));
35              videoView.start();
36          } else {
37              videoView.getBackground().setAlpha(0);           //将背景设为透明
38          }
39      }
40      /**
41       * 创建弹幕解析器
42       */
43      private BaseDanmakuParser parser=new BaseDanmakuParser() {
44          @Override
45          protected IDanmakus parse() {
46              return new Danmakus();
47          }
48      };
49      /**
50       * 初始化弹幕
51       */
52      private void initDanmaku() {
53          danmakuView.setCallback(new DrawHandler.Callback() {    //设置回调函数
54              @Override
55              public void prepared() {
56                  showDanmaku=true;
57                  danmakuView.start();  //开始弹幕
58                  generateDanmakus();   //调用随机生成弹幕方法
59              }
60              @Override
61              public void updateTimer(DanmakuTimer timer) {
62              }
```

```
63          @Override
64          public void drawingFinished() {
65          }
66      });
67      danmakuContext=DanmakuContext.create();
68      danmakuView.enableDanmakuDrawingCache(true);                //提升屏幕绘制效率
69      danmakuView.prepare(parser, danmakuContext);                //进行弹幕准备
70      //为 danmakuView 设置点击事件
71      danmakuView.setOnClickListener(new View.OnClickListener() {
72          @Override
73          public void onClick(View view) {
74              if(sendLayout.getVisibility()==View.GONE) {
75                  sendLayout.setVisibility(View.VISIBLE);         //显示布局
76              } else {
77                  sendLayout.setVisibility(View.GONE);            //隐藏布局
78              }
79          }
80      });
81      //为"发送"按钮设置点击事件
82      sendButton.setOnClickListener(new View.OnClickListener() {
83          @Override
84          public void onClick(View view) {
85              String content=editText.getText().toString();
86              if(!TextUtils.isEmpty(content)) {
87                  addDanmaku(content, true);                      //添加一条弹幕
88                  editText.setText("");
89              }
90          }
91      });
92  }
93  /**
94   * 添加一条弹幕
95   * @param content    弹幕的具体内容
96   * @param border 弹幕是否有边框
97   */
98  private void addDanmaku(String content, boolean border) {
99      //创建弹幕对象,TYPE_SCROLL_RL 表示从右向左滚动的弹幕
100     BaseDanmaku danmaku=danmakuContext.mDanmakuFactory.createDanmaku(
101                     BaseDanmaku.TYPE_SCROLL_RL);
102     danmaku.text=content;
103     danmaku.padding=6;
104     danmaku.textSize=30;
105     danmaku.textColor=Color.WHITE;                  //弹幕文字的颜色
106     danmaku.setTime(danmakuView.getCurrentTime());
107     if(border) {
108         danmaku.borderColor=Color.BLUE;             //弹幕文字边框的颜色
109     }
110     danmakuView.addDanmaku(danmaku);                //添加一条弹幕
111 }
```

```
112     /**
113      *  随机生成一些弹幕内容
114      */
115     private void generateDanmakus() {
116         new Thread(new Runnable() {
117             @Override
118             public void run() {
119                 while (showDanmaku) {
120                     int num=new Random().nextInt(300);
121                     String content=""+num;
122                     addDanmaku(content, false);
123                     try {
124                         Thread.sleep(num);
125                     } catch(InterruptedException e) {
126                         e.printStackTrace();
127                     }
128                 }
129             }
130         }).start();
131     }
132     @Override
133     protected void onPause() {
134         super.onPause();
135         if(danmakuView !=null && danmakuView.isPrepared()) {
136             danmakuView.pause();
137         }
138     }
139     @Override
140     protected void onResume() {
141         super.onResume();
142         if(danmakuView !=null && danmakuView.isPrepared() &&
143             danmakuView.isPaused()) {
144             danmakuView.resume();
145         }
146     }
147     @Override
148     protected void onDestroy() {
149         super.onDestroy();
150         showDanmaku=false;
151         if(danmakuView !=null) {
152             danmakuView.release();
153             danmakuView=null;
154         }
155     }
156 }
```

(5) 修改清单文件 (AndroidManifest.xml)。由于本项目需要使用自己的图标,并且在项目创建时默认带有的标题栏不够美观,因此需要修改清单文件中＜application＞标签中的 icon 与 theme 属性,具体代码如下:

```
android:icon="@mipmap/barrage_icon"
android:theme="@style/Theme.AppCompat.Light.NoActionBar"
```

由于只有视频播放界面处于横屏时视频与弹幕才会更加清晰,因此需要设置 MainActivity 对应的＜activity＞标签中的 screenOrientation 属性为横屏,具体代码如下:

```
android:screenOrientation="landscape"
```

1.2 本章小结

本章主要讲解了一个弹幕程序,无论是逻辑还是代码都比较简单,可以让读者先练练手,后续 3 章将会继续采用这种小项目的形式,逐渐深入,从第 5 章开始实践企业级项目。

【思考题】

1. 如何设置弹幕界面?
2. 如何把弹幕信息显示到界面上?

第 2 章 VR

学习目标

- 掌握 VR 主界面的开发,实现主界面的跳转功能
- 掌握 VR 全景图片的开发,实现全景图片预览功能
- 掌握 VR 全景视频的开发,实现全景视频播放功能

最近总是听到大家讨论 VR,VR 主题乐园也是随处可见,那么究竟什么是 VR 呢？VR 是 Virtual Reality 的缩写,称为虚拟现实。虚拟现实是一种可以创建和体验虚拟世界的计算机仿真系统,它利用计算机生成虚拟环境,这种环境可以将多源信息相融合,拥有交互式的三维动态视景和实体行为,使用户沉浸到该环境中,从而体验不一样的乐趣。本章将通过一个小项目开启 VR 的神奇之旅。让我们一起踏上旅程吧！

2.1 全景图片与全景视频介绍

思政材料 2

使用 VR 眼镜所看到的虚拟环境实际上离不开全景图片和全景视频,正因为这些图片和视频都是全景的,因此在移动 VR 眼镜时才会呈现出三维立体效果,使人沉浸在虚拟环境中。下面一起看一看全景图片和全景视频的效果,如图 2-1 所示。

从图 2-1 中可以看出,全景图片和全景视频都分为上下两部分,全景图片的上部分通过 VR 眼镜投射到左眼,下部分投射到右眼,全景视频也是同样的道理。

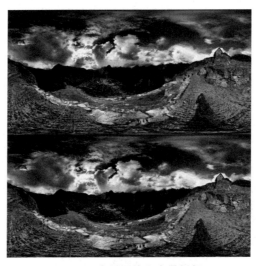

图 2-1 全景图片和全景视频

图 2-1 （续）

2.2 VR 主界面

任务综述

在 VR 项目的主界面中，包含一个 VR 全景图片按钮与一个 VR 全景视频按钮，点击这两个按钮会分别跳转到对应的界面。

【任务 2-1】 VR 主界面

【任务分析】

VR 项目的主界面效果如图 2-2 所示。

图 2-2 VR 项目的主界面

【任务实施】

（1）创建项目。首先创建一个项目，将其命名为 VR，指定包名为 com.itheima.vr。

（2）导入界面图片。在 Android Studio 中，切换到 Project 选项卡，在 res 文件夹中创建一个 drawable-hdpi 文件夹，将主界面所需要的图片 vr_bg.png、vr_pic.png、vr_video.png 导入 drawable-hdpi 文件夹，将项目的 icon 图标 vr_icon.png 导入 mipmap-hdpi 文件夹。

（3）放置界面控件。在 activity_main.xml 文件中，放置 2 个 ImageView 控件分别用于显示"VR 图片"按钮与"VR 视频"按钮，具体代码如文件 2-1 所示。

【文件 2-1】 activity_main.xml

```xml
1  <?xml version="1.0" encoding="utf-8"?>
2  <LinearLayout xmlns:android="http://schemas.android.com/apk/res/android"
3      android:id="@+id/activity_main"
4      android:layout_width="match_parent"
5      android:layout_height="match_parent"
6      android:background="@drawable/vr_bg"
7      android:gravity="center"
8      android:orientation="vertical">
9      <ImageView
10         android:id="@+id/iv_pic"
11         android:layout_width="200dp"
12         android:layout_height="200dp"
13         android:src="@drawable/vr_pic" />
14     <ImageView
15         android:id="@+id/iv_video"
16         android:layout_width="200dp"
17         android:layout_height="200dp"
18         android:src="@drawable/vr_video" />
19 </LinearLayout>
```

【任务 2-2】 VR 主界面逻辑代码

【任务分析】

VR 主界面主要显示一个"全景图片"按钮与一个"全景视频"按钮，分别点击这两个按钮会跳转到对应的全景图片界面与全景视频界面。

【任务实施】

（1）获取界面控件。在 MainActivity 中创建界面控件的初始化方法 init()，用于获取 VR 主界面所要用到的控件。

（2）处理点击事件。将 MainActivity 类实现 OnClickListener 接口并重写 onClick() 方法，在该方法中获取全景图片与全景视频按钮的 Id，具体代码如文件 2-2 所示。

【文件 2-2】 MainActivity.java

```java
1  package com.itheima.vr;
2  public class MainActivity extends AppCompatActivity implements View.
```

```
3  OnClickListener {
4      private ImageView iv_pic, iv_video;
5      @Override
6      protected void onCreate(Bundle savedInstanceState) {
7          super.onCreate(savedInstanceState);
8          setContentView(R.layout.activity_main);
9          init();
10     }
11     private void init() {
12         iv_pic= (ImageView) findViewById(R.id.iv_pic);
13         iv_video= (ImageView) findViewById(R.id.iv_video);
14         iv_pic.setOnClickListener(this);
15         iv_video.setOnClickListener(this);
16     }
17     @Override
18     public void onClick(View v) {
19         switch(v.getId()) {
20             case R.id.iv_pic:
21                 //跳转到VR全景图片界面
22                 break;
23             case R.id.iv_video:
24                 //跳转到VR全景视频界面
25                 break;
26         }
27     }
28 }
```

2.3　VR 全景图片

任务综述

在程序加载全景图片时，一开始只显示一个屏幕，移动手机后会看到不同角度的图像。如果点击屏幕右下角的眼镜图标，就会提示"将您的手机放入 Cardboard 眼镜"，将手机横放后稍等片刻便会切换为两个屏幕，此时移动手机，左右眼就会看到不同的图像。

【任务 2-3】　VR 全景图片界面

【任务分析】

VR 项目的全景图片界面主要显示一张全景图片，点击屏幕右下角的眼镜图标会显示两个屏幕，供左右眼进行观看，界面效果如图 2-3 所示。

【任务实施】

（1）创建 VR 全景图片界面。在 com.itheima.vr 包中创建一个 Empty Activity 类，命名为 VRPicActivity 并将布局文件名指定为 activity_vrpic。

（2）导入全景图片。在 Android Studio 中，将选项卡切换到 Project 选项，在 app/src/

 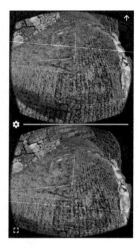

图 2-3　全景图片界面

main 文件夹下创建一个 assets 文件夹，将全景图片 andes.jpg 导入该文件夹。

（3）添加 VR 项目需要的库（common、commonwidget、panowidget、videowidget）。在 Android Studio 中，选择 File→New→Import Module 选项把 common 库导入项目，选中项目，右击选择 Open Module Settings 选项，然后选择 Dependencies 选项卡，单击右上角的绿色加号并选择 Module Dependency 选项，把 common 库加入项目。其他 3 个库文件的添加方式与 common 库相同，读者自行添加即可。

注意：全景图片依赖的库为 common、commonwidget、panowidget，全景视频依赖的库为 common、commonwidget、videowidget，此处为方便起见，可以将其一起导入项目。

（4）放置界面控件。在布局文件中，放置一个 VrPanoramaView 控件用于加载全景图片，具体代码如文件 2-3 所示。

【文件 2-3】 activity_vrpic.xml

```xml
1  <?xml version="1.0" encoding="utf-8"?>
2  <FrameLayout xmlns:android="http://schemas.android.com/apk/res/android"
3      android:layout_width="match_parent"
4      android:layout_height="match_parent">
5      <com.google.vr.sdk.widgets.pano.VrPanoramaView
6          android:id="@+id/vr_image"
7          android:layout_width="match_parent"
8          android:layout_height="match_parent" />
9  </FrameLayout>
```

（5）修改清单文件。由于本项目使用的是自己的图标，因此需要设置清单文件中＜application＞标签的 icon 属性为@mipmap/vr_icon。

由于项目创建时默认带有的标题栏不够美观，因此需要设置清单文件中＜application＞标签的 theme 属性为@style/Theme.AppCompat.Light.NoActionBar，删除标题栏。

由于本项目需要加载较大的图片和视频，因此需要设置清单文件中＜application＞标签的 largeHeap 属性为 true，为程序提供更大的内存空间，避免出现 OOM。

【任务 2-4】 VR 全景图片界面逻辑代码

【任务分析】

VR 全景图片界面的逻辑代码实际上只加载一张全景图片,并在图片加载完成后设置图片的立体显示效果,其界面上的眼镜图标是 VrPanoramaView 控件自带的,不需要特意设置。

【任务实施】

(1) 创建 init() 方法。在 VRPicActivity 中创建界面控件的初始化方法 init(),在该方法中获取 VrPanoramaView 控件并异步加载 assets 文件夹中的图片 andes.jpg。

(2) 异步加载全景图片。在 init() 方法中的异步加载全景图片任务中,重写 doInBackground() 方法和 onPostExecute() 方法,其中 doInBackground() 方法主要用于从 assets 文件夹中加载图片,onPostExecute() 方法主要用于把加载的图片显示到界面上,具体代码如文件 2-4 所示。

【文件 2-4】 VRPicActivity.java

```
1   package com.itheima.vr;
2   public class VRPicActivity extends AppCompatActivity {
3       private VrPanoramaView vrImage;
4       private AsyncTask<Void, Void, Bitmap> task;
5       @Override
6       protected void onCreate(Bundle savedInstanceState) {
7           super.onCreate(savedInstanceState);
8           setContentView(R.layout.activity_vrpic);
9           init();
10      }
11      private void init() {
12          vrImage= (VrPanoramaView) findViewById(R.id.vr_image);
13          vrImage.setEventListener(new VrPanoramaEventListener());
14          //异步加载全景图片
15          task=new AsyncTask<Void, Void, Bitmap>() {
16              @Override
17              protected Bitmap doInBackground(Void... params) {
18                  try {
19                      //从 assets 目录中加载图片
20                      InputStream is=getAssets().open("andes.jpg");
21                      Bitmap bitmap=BitmapFactory.decodeStream(is);
22                      is.close();
23                      return bitmap;
24                  } catch(Exception e) {
25                      e.printStackTrace();
26                  }
27                  return null;
28              }
29              @Override
30              protected void onPostExecute(Bitmap bitmap) {
```

```
31              super.onPostExecute(bitmap);
32              if(bitmap !=null) {
33                  //加载配置
34                  VrPanoramaView.Options options=new VrPanoramaView.Options();
35                  //设置图片为立体效果
36                  options.inputType=VrPanoramaView.Options.TYPE_STEREO_OVER_UNDER;
37                  //加载图片
38                  vrImage.loadImageFromBitmap(bitmap, options);
39              }
40          }
41      }.execute();
42  }
43  @Override
44  protected void onResume() {
45      super.onResume();
46      vrImage.resumeRendering();           //恢复渲染
47  }
48  @Override
49  protected void onPause() {
50      super.onPause();
51      vrImage.pauseRendering();            //暂停渲染
52  }
53  @Override
54  protected void onDestroy() {
55      super.onDestroy();
56      vrImage.shutdown();                  //关闭渲染释放内存
57      if(task !=null) {
58          task.cancel(true);               //停止异步任务
59          task=null;
60      }
61  }
62 }
```

（3）修改 MainActivity.java 文件。由于在点击主界面上的"VR 图片"按钮时会跳转到 VR 全景图片界面，因此需要找到文件 2-2 中的 onClick()方法，在该方法中的"//跳转到 VR 全景图片界面"语句下方添加如下代码：

```
Intent pic=new Intent(MainActivity.this,VRPicActivity.class);
startActivity(pic);
```

2.4 VR 全景视频

任务综述

全景视频其实和全景图片的原理相同，也分为上下两部分，上部分投射到左眼，下部分投射到右眼，戴上 VR 眼镜后就可以观看全景视频了。

【任务 2-5】 VR 全景视频界面

【任务分析】

全景视频界面与全景图片界面类似,点击右下角的眼镜图标后会显示两个屏幕,供左右眼同时观看全景视频,界面效果如图 2-4 所示。

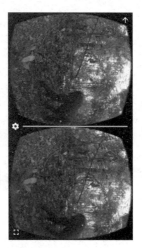

图 2-4 全景视频界面

【任务实施】

(1) 创建 VR 全景视频界面。在 com.itheima.vr 包中创建一个 Empty Activity 类,命名为 VRVideoActivity 并将布局文件名指定为 activity_vrvideo。

(2) 导入全景视频。将全景视频界面播放的视频 congo.mp4 导入 assets 文件夹。

(3) 放置界面控件。在布局文件中,放置一个 VrVideoView 控件用于显示全景视频,一个 SeekBar 控件用于显示视频播放的进度条,一个 TextView 控件用于显示视频播放进度的百分比,具体代码如文件 2-5 所示。

【文件 2-5】 activity_vrvideo.xml

```
1   <?xml version="1.0" encoding="utf-8"?>
2   <FrameLayout xmlns:android="http://schemas.android.com/apk/res/android"
3       android:layout_width="match_parent"
4       android:layout_height="match_parent">
5       <com.google.vr.sdk.widgets.video.VrVideoView
6           android:id="@+id/vr_video"
7           android:layout_width="match_parent"
8           android:layout_height="match_parent" />
9       <LinearLayout
10          android:layout_width="match_parent"
11          android:layout_height="wrap_content"
12          android:background="#FFF000"
13          android:orientation="horizontal">
```

```
14      <SeekBar
15          android:id="@+id/seekbar"
16          android:layout_width="0dp"
17          android:layout_height="wrap_content"
18          android:layout_gravity="center_vertical"
19          android:layout_weight="1" />
20      <TextView
21          android:id="@+id/seekbar_text"
22          android:layout_width="wrap_content"
23          android:layout_height="wrap_content"
24          android:layout_gravity="center_vertical"
25          android:padding="20dp"
26          android:text="--:--"
27          android:textSize="19sp" />
28  </LinearLayout>
29 </FrameLayout>
```

【任务 2-6】 VR 全景视频界面逻辑代码

【任务分析】

VR 全景视频界面的逻辑代码与全景图片界面类似,主要是在异步任务中加载全景视频,并设置视频的加载效果为 3D 效果,然后在界面上播放视频,并显示视频的播放进度。

【任务实施】

(1) 创建 init() 方法。在 VRVideoActivity 中创建一个 init() 方法,在该方法中调用 setOnSeekBarChangeListener() 方法用于监听视频播放的进度条,调用 setEventListener() 方法用于监听视频的暂停与播放,调用 AsyncTask 类异步加载 assets 文件夹中的视频,具体代码如文件 2-6 所示。

【文件 2-6】 VRVideoActivity.java

```
1   package com.itheima.vr;
2   public class VRVideoActivity extends AppCompatActivity {
3       private VrVideoView videoView;
4       private AsyncTask<Void, Void, Void> task;
5       private TextView seekbarText;
6       private SeekBar seekbar;
7       @Override
8       protected void onCreate(Bundle savedInstanceState) {
9           super.onCreate(savedInstanceState);
10          setContentView(R.layout.activity_vrvideo);
11          init();
12      }
13      private void init() {
14          seekbar= (SeekBar) findViewById(R.id.seekbar);
15          seekbarText= (TextView) findViewById(R.id.seekbar_text);
16          seekbar.setOnSeekBarChangeListener(new SeekBar.OnSeekBarChangeListener() {
```

```java
17          @Override
18          public void onProgressChanged(SeekBar seekBar, int progress,
19          boolean fromUser)
20          {
21              if(fromUser) {
22                  long duration=videoView.getDuration();
23                  long newPosition= (long) (progress * 0.01f * duration);
24                  videoView.seekTo(newPosition);
25              }
26          }
27          @Override
28          public void onStartTrackingTouch(SeekBar seekBar) {
29          }
30          @Override
31          public void onStopTrackingTouch(SeekBar seekBar) {
32          }
33      });
34      videoView= (VrVideoView) findViewById(R.id.vr_video);
35      videoView.setTag(true);
36      videoView.setEventListener(new VrVideoEventListener() {
37          @Override
38          public void onClick() {
39              super.onClick();
40              boolean isPlay= (boolean) videoView.getTag();
41              if(isPlay) {
42                  isPlay=false;
43                  videoView.pauseVideo();           //暂停播放
44              } else {
45                  isPlay=true;
46                  videoView.playVideo();            //继续播放
47              }
48              videoView.setTag(isPlay);
49          }
50          //视图画面切换到下一帧时被调用
51          @Override
52          public void onNewFrame() {
53              super.onNewFrame();
54              seekbar.setMax(100);
55              long duration=videoView.getDuration();             //获取视频时长,单位为毫秒
56              long currentPosition=videoView.getCurrentPosition();     //获取当前位置
57              int percent= (int) (currentPosition * 100f / duration+0.5f);
58              seekbar.setProgress(percent);
59              seekbarText.setText(percent+"% ");
60          }
61          @Override
62          public void onLoadSuccess() {
63              super.onLoadSuccess();
64              Toast.makeText(VRVideoActivity.this, "开始播放", Toast.LENGTH_SHORT)
65                      .show();
66              seekbar.setMax(100);
```

```
67             seekbar.setProgress(0);
68             seekbarText.setText("0% ");
69         }
70         @Override
71         public void onLoadError(String errorMessage) {
72             super.onLoadError(errorMessage);
73             Toast.makeText(VRVideoActivity.this, "播放出错", Toast.LENGTH_SHORT)
74                     .show();
75         }
76         @Override
77         public void onCompletion() {
78             super.onCompletion();
79             seekbar.setMax(100);
80             seekbar.setProgress(100);
81             seekbarText.setText("100% ");
82         }
83     });
84     //创建异步任务
85     task=new AsyncTask<Void, Void, Void>() {
86         @Override
87         protected Void doInBackground(Void... params) {
88             String fileName="congo.mp4";
89             VrVideoView.Options option=new VrVideoView.Options();
90             //非流媒体
91             option.inputFormat=VrVideoView.Options.FORMAT_DEFAULT;
92             //设置 3D 效果
93             option.inputType=VrVideoView.Options.TYPE_STEREO_OVER_UNDER;
94             try {
95                 //从资产目录加载视频
96                 videoView.loadVideoFromAsset(fileName, option);
97             } catch(Exception e) {
98                 e.printStackTrace();
99             }
100            return null;
101        }
102        @Override
103        protected void onPostExecute(Void aVoid) {
104            super.onPostExecute(aVoid);
105            Toast.makeText(VRVideoActivity.this, "加载成功开始播放",
106                    Toast.LENGTH_SHORT).show();
107        }
108    }.execute();
109 }
110 //细节一,切换模式造成屏幕变黑
111 @Override
112 protected void onResume() {
113     super.onResume();
114     videoView.resumeRendering();         //重新渲染
115 }
116 @Override
```

```
117    protected void onPause() {
118        super.onPause();
119        videoView.pauseRendering();          //停止渲染
120    }
121    //细节二,AsyncTask 处理
122    @Override
123    protected void onDestroy() {
124        super.onDestroy();
125        videoView.shutdown();                //停止
126        if(task !=null) {
127            task.cancel(true);
128            task=null;
129        }
130    }
131 }
```

（2）修改 MainActivity.java 文件。由于在点击主界面上的"VR 视频"按钮时会跳转到 VR 全景视频界面，因此需要找到文件 2-2，在该文件的 onClick()方法中的"//跳转到 VR 全景视频界面"语句下方添加如下代码：

```
Intent video=new Intent(MainActivity.this,VRVideoActivity.class);
startActivity(video);
```

2.5　本章小结

本章主要讲解了 VR 技术的应用，核心内容是 VR 全景图片与 VR 全景视频。VR 技术是未来科研发展的一个方向，掌握 VR 技术可以在当今市场中更具竞争力，因此建议读者亲手练习该项目。

【思考题】

1. 如何加载全景图片？
2. 如何播放全景视频？

第 3 章

图 表 库

学习目标

- 掌握线形图的使用,使用线形图展示天气信息
- 掌握饼状图的使用,使用饼状图展示教育领域投资情况
- 掌握柱状图的使用,使用柱状图展示互联网教育市场的规模

实际开发中经常会使用第三方的图表库展示数据,其中 HelloCharts 就是一款非常优秀的开源图表库。相比其他图表库来说,HelloCharts 的功能更加完善、界面更加美观、坐标更加准确,并且 HelloCharts 还支持图表缩放、平移等特性。本章将带领读者进一步学习 HelloCharts 的使用,制作美观的界面统计图。

3.1 线形图

思政材料 3

任务综述

线形图是通过 LineChartView 控件实现的。线形图通常分为两种情况,一种是通过直线连接各个数据点,另一种是通过平滑的曲线连接各个数据点。本项目通过一张直线连接的线形图展示一周的温度变化。

【任务 3-1】 线形图界面

【任务分析】

线形图的界面主要由横坐标、纵坐标以及线形图组成,其中横坐标显示周一到周日,纵坐标显示温度(24.0~26.8℃),线形图是根据每天的温度坐标显示的,界面效果如图 3-1 所示。

【任务实施】

(1) 创建项目。首先创建一个项目,将其命名为 HelloCharts,指定包名为 com.itheima.hellocharts。将项目的 icon 图标 chart_icon.png 导入 mipmap-hdpi 文件夹。

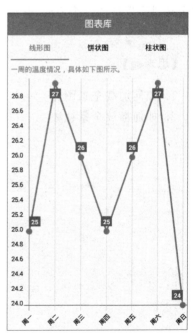

图 3-1 线形图界面

（2）添加 design 库。由于图表库界面中用到了 design 库中的 TabLayout 类，因此需要选中项目，右击选择 Open Module Settings 选项后选择 Dependencies 选项卡，单击右上角的绿色加号并选择 Library dependency 选项，将 com.android.support:design:25.3.1 库添加到项目中。

（3）添加 hellocharts-library 库。由于图表库界面中的线形图、饼状图以及柱状图都用到了 hellocharts-library 库中的控件，因此需要在 Android Studio 中选择 File→New→Import Module 选项，把第三方库 hellocharts-library 导入项目中，接着选中项目，右击选择 Open Module Settings 选项，选择 Dependencies 选项卡，单击右上角的绿色加号并选择 Module Dependency 选项，把 hellocharts-library 库添加到主项目。

（4）放置界面控件。在 activity_main.xml 文件中，放置一个 TextView 控件用于显示标题，一个 ViewPager 控件用于加载 3 个图形界面，一个 TabLayout 控件用于加载顶部文本与 ViewPager 控件中的内容，具体代码如文件 3-1 所示。

【文件 3-1】 activity_main.xml

```xml
1  <?xml version="1.0" encoding="utf-8"?>
2  <LinearLayout xmlns:android="http://schemas.android.com/apk/res/android"
3      xmlns:app="http://schemas.android.com/apk/res-auto"
4      android:layout_width="fill_parent"
5      android:layout_height="fill_parent"
6      android:orientation="vertical">
7      <TextView
8          android:id="@+id/tv_main_title"
9          android:layout_width="match_parent"
10         android:layout_height="45dp"
11         android:background="@android:color/holo_blue_light"
12         android:gravity="center"
13         android:text="图表库"
14         android:textColor="@android:color/white"
15         android:textSize="18sp" />
16     <android.support.design.widget.TabLayout
17         android:id="@+id/tabs"
18         android:layout_width="match_parent"
19         android:layout_height="wrap_content"
20         android:layout_marginLeft="8dp"
21         android:layout_marginRight="8dp"
22         app:tabIndicatorColor="@android:color/holo_blue_light"
23         app:tabSelectedTextColor="@android:color/holo_blue_light"
24         app:tabTextColor="@android:color/black" />
25     <!--可滑动的布局内容-->
26     <android.support.v4.view.ViewPager
27         android:id="@+id/vp_view"
28         android:layout_width="match_parent"
29         android:layout_height="wrap_content" />
30 </LinearLayout>
```

（5）创建线形图界面。在 res/layout 文件夹中创建一个布局文件 layout_line_chart.xml 文件。

（6）放置界面控件。在 layout_line_chart.xml 文件中，放置一个 TextView 控件用于显示线形图的介绍信息，一个 LineChartView 控件用于显示线形图，具体代码如文件 3-2 所示。

【文件 3-2】 layout_line_chart.xml

```xml
1  <?xml version="1.0" encoding="utf-8"?>
2  <LinearLayout xmlns:android="http://schemas.android.com/apk/res/android"
3      android:id="@+id/activity_line"
4      android:layout_width="match_parent"
5      android:layout_height="match_parent"
6      android:orientation="vertical">
7      <TextView
8          android:layout_width="match_parent"
9          android:layout_height="wrap_content"
10         android:gravity="center_vertical"
11         android:padding="8dp"
12         android:text="一周的温度情况,具体如下图所示。"
13         android:textColor="@android:color/darker_gray"
14         android:textSize="14sp" />
15     <lecho.lib.hellocharts.view.LineChartView
16         android:id="@+id/lv_chart"
17         android:layout_width="wrap_content"
18         android:layout_height="480dp" />
19 </LinearLayout>
```

（7）修改清单文件。由于本项目使用的是自己的图标，因此需要设置清单文件中 <application> 标签的 icon 属性为 @mipmap/chart_icon。

由于项目创建时默认带有的标题栏不够美观，因此需要设置清单文件中 <application> 标签中的 theme 属性为 @style/Theme.AppCompat.Light.NoActionBar，删除标题栏。

【任务 3-2】 创建 ViewPagerAdapter

【任务分析】

由于在 activity_main.xml 文件中添加一个 ViewPager 控件，用于加载线形图、饼状图以及柱状图的 3 个页面并且使其可左右滑动，因此需要创建一个 ViewPagerAdapter 对 ViewPager 控件进行数据填充。

【任务实施】

在 com.itheima.hellocharts 包中创建一个 ViewPagerAdapter 类继承 PagerAdapter 类，并重写 getCount() 方法、isViewFromObject() 方法、instantiateItem() 方法、destroyItem() 方法以及 getPageTitle() 方法，这些方法分别用于获取页卡数、判断容器中的 View 是否与一个 key 值相关联、添加页卡、删除页卡以及页卡标题。具体代码如文件 3-3 所示。

【文件 3-3】 ViewPagerAdapter.java

```java
package com.itheima.hellocharts;
public class ViewPagerAdapter extends PagerAdapter {
    private List<View>mViewList;
    private List<String>mTitleList;
    public ViewPagerAdapter(List<View>mViewList, List<String>mTitleList) {
        this.mViewList=mViewList;
        this.mTitleList=mTitleList;
    }
    @Override
    public int getCount() {
        return mViewList.size();          //页卡数
    }
    @Override
    public boolean isViewFromObject(View view, Object object) {
        return view==object;              //官方推荐写法
    }
    @Override
    public Object instantiateItem(ViewGroup container, int position) {
        container.addView(mViewList.get(position));   //添加页卡
        return mViewList.get(position);
    }
    @Override
    public void destroyItem(ViewGroup container, int position, Object object) {
        container.removeView(mViewList.get(position));   //删除页卡
    }
    @Override
    public CharSequence getPageTitle(int position) {
        return mTitleList.get(position);   //页卡标题
    }
}
```

【任务 3-3】 线形图界面逻辑代码

【任务分析】

线形图界面主要是根据已知的 X 轴数据以及一周的温度数据显示一周温度波动的线形图。

【任务实施】

（1）获取界面控件。在 MainActivity 中创建界面控件的初始化方法 init()，用于获取线形图界面所要用到的控件。

（2）设置 X 轴标注。在 MainActivity 中创建一个 setAxisXLables() 方法，用于把周一到周日的日期数据存放到 X 轴的数据集合中。

（3）设置线形图中的数据点。在 MainActivity 中创建一个 setAxisPoints() 方法，用于把一周的温度数据存放到所有点的数据集合中。

（4）初始化线形图。在 MainActivity 中创建一个 initLineChart() 方法，在该方法中分别设置 X 轴、Y 轴以及线形图的属性，具体代码如文件 3-4 所示。

【文件 3-4】 MainActivity.java

```java
1   package com.itheima.hellocharts;
2   public class MainActivity extends AppCompatActivity {
3       private TabLayout mTabLayout;
4       private ViewPager mViewPager;
5       private LayoutInflater mInflater;
6       private View view1;
7       private List<String>mTitleList=new ArrayList<>();         //页卡标题集合
8       private List<View>mViewList=new ArrayList<>();            //页卡视图集合
9       //线形图
10      private int[] temperature={25, 27, 26, 25, 26, 27, 24};   //图表的数据点
11      //X轴的标注
12      private String[] lineData={"周一","周二","周三","周四",
13                                  "周五","周六","周日"};
14      private List<PointValue>pointValues=new ArrayList<PointValue>();
15      private List<AxisValue>axisValues=new ArrayList<AxisValue>();
16      private LineChartView lineChartView;
17      @Override
18      protected void onCreate(Bundle savedInstanceState) {
19          super.onCreate(savedInstanceState);
20          setContentView(R.layout.activity_main);
21          init();
22      }
23      private void init() {
24          mViewPager=(ViewPager) findViewById(R.id.vp_view);
25          mTabLayout=(TabLayout) findViewById(R.id.tabs);
26          mInflater=LayoutInflater.from(this);
27          view1=mInflater.inflate(R.layout.layout_line_chart, null);
28          //添加页卡视图
29          mViewList.add(view1);
30          //添加页卡标题
31          mTitleList.add("线形图");
32          mTitleList.add("饼状图");
33          mTitleList.add("柱状图");
34          //设置 Tab 模式,当前为系统默认模式
35          mTabLayout.setTabMode(TabLayout.MODE_FIXED);
36          //添加选项卡
37          mTabLayout.addTab(mTabLayout.newTab().setText(mTitleList.get(0)));
38          mTabLayout.addTab(mTabLayout.newTab().setText(mTitleList.get(1)));
39          mTabLayout.addTab(mTabLayout.newTab().setText(mTitleList.get(2)));
40          ViewPagerAdapter mAdapter=new ViewPagerAdapter(mViewList, mTitleList);
41          mViewPager.setAdapter(mAdapter);                      //给 ViewPager 设置适配器
42          mTabLayout.setupWithViewPager(mViewPager);            //给 TabLayout 设置关联
43          //线形图
44          lineChartView=(LineChartView) view1.findViewById(R.id.lv_chart);
45          setAxisXLables();                                     //获取 X 轴的标注
46          setAxisPoints();                                      //设置坐标点
```

```
47            initLineChart();               //初始化线形图
48        }
49        /**
50         * 设置X轴的标注
51         */
52        private void setAxisXLables() {
53            for(int i=0; i<lineData.length; i++) {
54                axisValues.add(new AxisValue(i).setLabel(lineData[i]));
55            }
56        }
57        /**
58         * 设置线形图中的每个数据点
59         */
60        private void setAxisPoints() {
61            for(int i=0; i<temperature.length; i++) {
62                pointValues.add(new PointValue(i, temperature[i]));
63            }
64        }
65        /**
66         * 初始化线形图
67         */
68        private void initLineChart() {
69            //设置线的颜色、形状等属性
70            Line line=new Line();
71            line.setColor(Color.parseColor("#33b5e5"));
72            line.setShape(ValueShape.CIRCLE);  //线形图上的数据点的形状为圆形
73            line.setCubic(false);              //曲线是否平滑,即是曲线还是折线
74            line.setHasLabels(true);           //曲线的数据坐标是否加上备注
75            line.setHasLines(true);            //是否显示线条,如果为false则没有曲线只显示点
76            line.setHasPoints(true);           //是否显示圆点,如果为false则没有圆点只显示线
77            line.setValues(pointValues);
78            List<Line> lines=new ArrayList<Line>();
79            lines.add(line);
80            LineChartData data=new LineChartData();
81            data.setLines(lines);
82            //X轴
83            Axis axisX=new Axis();
84            axisX.setHasTiltedLabels(true);    //X轴字体是斜体显示还是正体,true是斜体
85            axisX.setTextColor(Color.BLACK);   //设置字体颜色
86            axisX.setMaxLabelChars(5);         //设置坐标轴标签显示的最大字符数
87            axisX.setValues(axisValues);       //填充X轴的坐标名称
88            data.setAxisXBottom(axisX);        //设置X轴在底部
89            axisX.setHasLines(true);           //X轴分割线
90            //Y轴
91            Axis axisY=new Axis();
92            data.setAxisYLeft(axisY);          //设置Y轴在左侧
93            axisY.setTextColor(Color.BLACK);   //设置字体颜色
94            axisY.setMaxLabelChars(5);         //设置坐标轴标签显示的最大字符数
95            //设置线形图的行为属性,如支持缩放、滑动以及平移
96            lineChartView.setInteractive(true);
```

```
 97        lineChartView.setZoomType(ZoomType.HORIZONTAL);       //设置缩放类型为水平缩放
 98        lineChartView.setMaxZoom((float) 2);                  //最大放大比例
 99        lineChartView.setContainerScrollEnabled(true,
100                ContainerScrollType.HORIZONTAL);
101        lineChartView.setLineChartData(data);
102        lineChartView.setVisibility(View.VISIBLE);
103     }
104 }
```

3.2　饼状图

任务综述

饼状图是通过 PieChartView 控件实现的。饼状图大致分为两种类型，一种是实心的饼状图，另一种是空心的饼状图。本项目以一个空心饼状图展示 2017 年互联网教育细分领域的投资情况，当点击饼状图的某一扇形时，扇形会凸出显示，并且扇形中的数据以及数据占比会显示在空心圆中。

【任务 3-4】　饼状图界面

【任务分析】

饼状图界面主要是以 2017 年互联网教育细分领域的投资情况为数据显示 5 个不同颜色的扇形图，界面效果如图 3-2 所示。

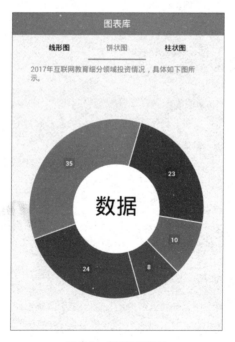

图 3-2　饼状图界面

【任务实施】

（1）创建饼状图界面。在 res/layout 文件夹中创建一个布局文件 layout_pie_chart.xml。

（2）放置界面控件。在布局文件中，放置一个 TextView 控件用于显示饼状图的介绍信息，一个 PieChartView 控件用于显示饼状图，具体代码如文件 3-5 所示。

【文件 3-5】　layout_pie_chart.xml

```
1   <?xml version="1.0" encoding="utf-8"?>
2   <LinearLayout xmlns:android="http://schemas.android.com/apk/res/android"
3       android:id="@+id/activity_main"
4       android:layout_width="match_parent"
5       android:layout_height="match_parent"
6       android:orientation="vertical">
7       <TextView
8           android:layout_width="match_parent"
9           android:layout_height="wrap_content"
10          android:gravity="center_vertical"
11          android:padding="8dp"
12          android:text="2017年互联网教育细分领域投资情况,具体如下图所示。"
13          android:textColor="@android:color/darker_gray"
14          android:textSize="14sp" />
15      <lecho.lib.hellocharts.view.PieChartView
16          android:id="@+id/pv_chart"
17          android:layout_width="wrap_content"
18          android:layout_height="wrap_content" />
19  </LinearLayout>
```

【任务 3-5】　饼状图界面逻辑代码

【任务分析】

饼状图界面主要是根据 2017 年互联网教育细分领域投资情况为数据，显示每个投资领域所占的数据以及百分比信息。点击饼状图中的任意一个扇形，扇形会凸出显示，并在空心圆中显示该领域的名称、占比以及具体数据。

【任务实施】

（1）创建饼状图所需字段与数据。由于饼状图界面需要获取饼状图控件并创建一些需要的字段与数据，因此在 MainActivity 中的"private LineChartView lineChartView;"语句下方添加如下代码：

```
1   //饼状图
2   private PieChartView pieChartView;
3   private PieChartData pieCharData;
4   private List<SliceValue> sliceValues=new ArrayList<SliceValue>();
5   private int[] pieData={8, 24, 35, 23, 10};        //饼状图中的数据
6   private int[] color={Color.parseColor("#356fb3"), Color.parseColor("#b53633"),
```

```
7     Color.parseColor("#86aa3d"), Color.parseColor("#6a4b90"), Color.
8     parseColor("#2e9cba")};              //饼状图每块的颜色
9     private String[] stateChar={"高等教育","职业教育","语言培训","K12教育","其他"};
```

把语句"private View view1;"改为语句"private View view1,view2;"。

（2）设置饼状图数据。在MainActivity中创建一个setPieChartData()方法用于设置饼状图数据到SliceValue集合中，具体代码如下所示：

```
1   /**
2    * 设置饼状图中的数据
3    */
4   private void setPieChartData() {
5       for(int i=0; i<pieData.length;++i) {
6           SliceValue sliceValue=new SliceValue((float) pieData[i], color[i]);
7           sliceValues.add(sliceValue);          //添加到集合中
8       }
9   }
```

（3）初始化饼状图。在MainActivity中创建一个initPieChart()方法，在该方法中设置饼状图的中心圆、环形颜色、默认文字、透明度以及饼状图数据等，具体代码如下所示：

```
1   /**
2    * 初始化饼状图
3    */
4   private void initPieChart() {
5       pieCharData=new PieChartData();
6       pieCharData.setHasLabels(true);                        //显示标签
7       pieCharData.setHasLabelsOnlyForSelected(false);        //不用点击显示占的百分比
8       pieCharData.setHasLabelsOutside(false);                //数据是否显示在饼图外侧
9       pieCharData.setValues(sliceValues);                    //填充数据
10      pieCharData.setCenterCircleColor(Color.WHITE);         //设置环形中间的颜色
11      pieCharData.setHasCenterCircle(true);                  //是否显示中心圆
12      pieCharData.setCenterCircleScale(0.5f);                //设置中心圆所占饼图的比例
13      pieCharData.setCenterText1("数据");                     //设置中心圆默认显示的文字
14      pieChartView.setPieChartData(pieCharData);             //为饼图设置数据
15      pieChartView.setValueSelectionEnabled(true);           //选择饼状图中的块会变大
16      pieChartView.setAlpha(0.9f);                           //设置透明度
17      pieChartView.setCircleFillRatio(1f);                   //设置饼图大小,占整个View的比例
18  }
```

（4）计算饼状图数据所占的百分比。在MainActivity中创建一个calPercent()方法，在该方法中计算饼状图中每个模块占全部数据的百分比，具体代码如下所示：

```
1   /**
2    * 数据所占的百分比
3    */
4   private String calPercent(int i) {
```

```
5       String result="";
6       int sum=0;
7       for(int j=0; j<pieData.length; j++) {
8           sum+=pieData[j];
9       }
10      result=String.format("% .2f", (float) pieData[i] * 100 / sum)+"% ";
11      return result;
12  }
```

(5) 创建饼状图的点击监听事件。在 MainActivity 中创建一个 PieChartOnValue-SelectListener 监听器,并在 onValueSelected()方法中设置点击饼状图后需要显示的对应数据,具体代码如下所示:

```
1   /**
2    * 饼状图的事件监听器
3    */
4   PieChartOnValueSelectListener selectListener=
5                                   new PieChartOnValueSelectListener() {
6       @Override
7       public void onValueDeselected() {
8       }
9       @Override
10      public void onValueSelected(int arg0, SliceValue value) {
11          //选择对应图形后,在中间部分显示相应信息
12          pieCharData.setCenterText1(stateChar[arg0]);       //中心圆中的第一文本
13          pieCharData.setCenterText2(value.getValue()+"("+calPercent(arg0)+")");
14      }
15  };
```

(6) 加载饼状图布局并调用以上添加的方法。在 MainActivity 的 init()方法中,加载饼状图布局、设置饼状图监听事件、调用设置饼状图数据以及初始化饼状图的方法。在 init() 方法中的"view1=mInflater.inflate(R.layout.layout_line_chart,null);"语句下方添加如下代码:

```
view2=mInflater.inflate(R.layout.layout_pie_chart, null);
```

在"mViewList.add(view1);"语句下方添加如下代码:

```
mViewList.add(view2);
```

在"initLineChart(); //初始化线形图"方法中添加如下代码:

```
//饼状图
pieChartView=(PieChartView) view2.findViewById(R.id.pv_chart);
pieChartView.setOnValueTouchListener(selectListener);       //为饼状图设置事件监听器
setPieChartData();
initPieChart();
```

3.3 柱状图

任务综述

柱状图是通过 ColumnChartView 控件实现的。本项目以一个柱状图展示 2013—2017 年互联网教育市场的规模，当点击柱状图中的每个条目时，条目会放大并显示条目对应的数据。

【任务 3-6】 柱状图界面

【任务分析】

柱状图界面主要是以 2013—2017 年互联网教育市场规模为数据，显示 5 个不同颜色的条目，界面效果如图 3-3 所示。

【任务实施】

（1）创建柱状图界面。在 res/layout 文件夹中创建一个布局文件 layout_column_chart.xml。

（2）放置界面控件。在布局文件中，放置一个 TextView 控件用于显示柱状图的介绍信息，一个 ColumnChartView 控件用于显示柱状图，具体代码如文件 3-6 所示。

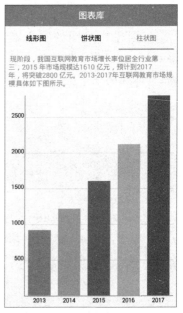

图 3-3 柱状图界面

【文件 3-6】 layout_column_chart.xml

```
1  <?xml version="1.0" encoding="utf-8"?>
2  <LinearLayout xmlns:android="http://schemas.android.com/apk/res/android"
3      android:layout_width="match_parent"
4      android:layout_height="match_parent"
5      android:orientation="vertical">
6      <TextView
7          android:layout_width="match_parent"
8          android:layout_height="wrap_content"
9          android:gravity="center_vertical"
10         android:padding="8dp"
11         android:text="现阶段,我国互联网教育市场增长率位居全行业第三,2015 年市场
12         规模达 1610 亿元,预计到 2017 年,将突破 2800 亿元。2013-2017 互联网教育市
13         场规模具体如下图所示。"
14         android:textColor="@android:color/darker_gray"
15         android:textSize="14sp" />
16     <lecho.lib.hellocharts.view.ColumnChartView
17         android:id="@+id/cv_chart"
18         android:layout_width="wrap_content"
19         android:layout_height="wrap_content"
```

```
20            android:layout_margin="8dp" />
21  </LinearLayout>
```

【任务 3-7】 柱状图界面逻辑代码

【任务分析】

柱状图界面主要是以不同颜色的条目显示从 2013—2017 年互联网教育市场规模的具体数据,每个不同颜色的条目显示一年的数据,点击柱状图中的任意一个条目,该条目会放大并显示对应的数据。

【任务实施】

(1) 创建柱状图所需字段与数据。由于柱状图界面需要获取柱状图控件并创建一些需要的字段与数据,因此在 MainActivity 中的"private String[] stateChar={"高等教育","职业教育","语言培训","K12 教育","其他"};"语句下方添加如下代码:

```
1  //柱状图
2  private String[] year=new String[]{"2013", "2014", "2015", "2016", "2017"};
3  private ColumnChartView columnChartView;
4  private ColumnChartData columnChartData;
5  private int[] columnY={500, 1000, 1500, 2000, 2500, 3000};
```

把语句"private View view1, view2;"改为语句"private View view1, view2, view3;"。

(2) 初始化柱状图。在 MainActivity 中创建一个 initColumnChart()方法,在该方法中设置柱状图的 X 轴标注、Y 轴标注、柱状图每个条目的颜色以及柱状图的数据等,具体代码如下所示:

```
1   /**
2    * 初始化柱状图
3    */
4   private void initColumnChart() {
5       List<AxisValue>axisValues=new ArrayList<AxisValue>();        //存储 X 轴标注
6       List<AxisValue>axisYValues=new ArrayList<AxisValue>();       //存储 Y 轴标注
7       List<Column>columns=new ArrayList<Column>();
8       List<SubcolumnValue> subcolumnValues;                        //存储
9       for(int k=0; k<columnY.length; k++) {
10          axisYValues.add(new AxisValue(k).setValue(columnY[k]));
11      }
12      for(int i=0; i<year.length;i++) {
13          subcolumnValues=new ArrayList<SubcolumnValue>();
14          for(int j=0; j<1;j++) {
15              switch(i+1) {
16                  case 1:
17                      subcolumnValues.add(new SubcolumnValue(924,
18                          ChartUtils.COLOR_BLUE));
19                      break;
```

```
20          case 2:
21              subcolumnValues.add(new SubcolumnValue(1220,
22                  ChartUtils.COLOR_GREEN));
23              break;
24          case 3:
25              subcolumnValues.add(new SubcolumnValue(1610,
26                  ChartUtils.COLOR_RED));
27              break;
28          case 4:
29              subcolumnValues.add(new SubcolumnValue(2125,
30                  ChartUtils.COLOR_ORANGE));
31              break;
32          case 5:
33              subcolumnValues.add(new SubcolumnValue(2805,
34                  ChartUtils.COLOR_VIOLET));
35              break;
36          }
37      }
38      //点击柱状图就可展示数据量
39      axisValues.add(new AxisValue(i).setLabel(year[i]));
40      columns.add(new Column(subcolumnValues).setHasLabelsOnlyForSelected(true));
41  }
42  //X轴
43  Axis axisX=new Axis(axisValues);
44  axisX.setHasLines(false);
45  axisX.setTextColor(Color.BLACK);
46  //Y轴
47  Axis axisY=new Axis(axisYValues);
48  axisY.setHasLines(true);                              //设置Y轴有线条显示
49  axisY.setTextColor(Color.BLACK);                      //设置文本颜色
50  axisY.setMaxLabelChars(5);                            //设置坐标轴标签显示的最大字符数
51                                                        //设置柱状图的相关属性
52  columnChartData=new ColumnChartData(columns);
53  columnChartData.setAxisXBottom(axisX);                //设置X轴在底部
54  columnChartData.setAxisYLeft(axisY);                  //设置Y轴在左侧
55  columnChartView.setColumnChartData(columnChartData);
56  columnChartView.setValueSelectionEnabled(true);       //设置柱状图可以被选择
57  columnChartView.setZoomType(ZoomType.HORIZONTAL);     //设置缩放类型为水平缩放
58  }
```

（3）加载柱状图布局并调用柱状图初始化方法。在 MainActivity 的 init()方法中,加载柱状图布局、初始化柱状图。在 init()方法中的"view2 = mInflater. inflate(R. layout. layout_pie_chart, null);"语句下方添加如下代码：

```
view3=mInflater.inflate(R.layout.layout_column_chart, null);
```

在"mViewList.add(view2);"语句下方添加如下代码：

```
mViewList.add(view3);
```

在"initPieChart();//初始化饼状图"方法中添加如下代码:

```
//柱状图
columnChartView= (ColumnChartView) view3.findViewById(R.id.cv_chart);
initColumnChart();
```

3.4 本章小结

本章主要针对饼状图、柱状图以及线形图进行了详细讲解,这三种图表在实际开发中比较常用,读者在学习过程中可以在此基础上继续探索其他图表。

【思考题】

1. 如何使用饼状图展示数据?
2. 如何使用柱状图展示数据?

第 4 章 3D 相册

学习目标
- 掌握"相册"界面的开发,实现相册的立体与倒影效果

现实生活中,人们浏览照片时,不同的相册软件有不同的浏览效果。本章主要实现一个3D效果的相册,该相册中的图片可以循环滑动展示。

4.1 相册

任务综述

思政材料 4

"相册"界面主要展示一组图片的立体效果,并显示每个图片的倒影,左右滑动可以切换不同的图片到界面的中间位置,同时在图片下方显示当前图片的标题,点击界面上的任意一个图片时,会弹出该图片的标题信息。

【任务 4-1】 "相册"界面

【任务分析】

"相册"界面由一组图片与一个图片标题组成,界面效果如图 4-1 所示。

图 4-1 3D 相册界面

【任务实施】

（1）创建项目。首先创建一个项目，将其命名为3DAlbum，指定包名为com.itheima.album，Activity名为AlbumActivity，对应布局名为activity_album。

注意：本项目的build.gradle文件中的minSdkVersion的值设置为17。

（2）导入界面图片。在Android Studio中，切换到Project选项卡，在res文件夹中创建一个drawable-hdpi文件夹，将"相册"界面所需要的图片img_1.png、img_2.png、img_3.png、img_4.png、img_5.png导入drawable-hdpi文件夹，将项目的icon图标album_icon.png导入mipmap-hdpi文件夹。

（3）引入library库。3D相册通过引入第三方库library中的FeatureCoverFlow控件显示。在Android Studio中，选择File→New→Import Module选项，把第三方库library导入项目，选中项目后右击选择Open Module Settings选项，然后选择Dependencies选项卡，单击右上角的绿色加号并选择Module Dependency选项，把library库加入主项目，library库的详细信息如图4-2所示。

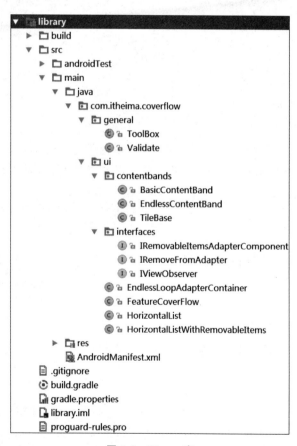

图4-2　library库

（4）放置界面控件。在布局文件中，放置一个自定义控件FeatureCoverFlow用于显示相册，一个TextSwitcher控件用于显示图片的标题，具体代码如文件4-1所示。

【文件 4-1】 activity_album.xml

```
1   <RelativeLayout xmlns:android="http://schemas.android.com/apk/res/android"
2       xmlns:coverflow="http://schemas.android.com/apk/res-auto"
3       android:layout_width="match_parent"
4       android:layout_height="match_parent"
5       android:background="@android:color/darker_gray">
6       <com.itheima.coverflow.ui.FeatureCoverFlow
7           android:id="@+id/fcf_coverflow"
8           android:layout_width="match_parent"
9           android:layout_height="match_parent"
10          coverflow:coverHeight="@dimen/album_height"
11          coverflow:coverWidth="@dimen/album_width"
12          coverflow:maxScaleFactor="1.5"
13          coverflow:reflectionGap="0dp"
14          coverflow:rotationThreshold="0.5"
15          coverflow:scalingThreshold="0.5"
16          coverflow:spacing="0.6" />
17      <TextSwitcher
18          android:id="@+id/ts_title"
19          android:layout_width="match_parent"
20          android:layout_height="wrap_content"
21          android:layout_alignParentBottom="true"
22          android:layout_centerVertical="true"
23          android:paddingBottom="16dp" />
24  </RelativeLayout>
```

（5）修改 dimens.xml 文件。由于"相册"界面中相册控件的宽和高是固定的，因此需要在 res/values 文件夹中的 dimens.xml 文件中添加相册的宽和高，具体代码如下所示。

```
<dimen name="album_width">120dp</dimen>
<dimen name="album_height">180dp</dimen>
```

【任务 4-2】 "相册"界面 Item

【任务分析】

由于"相册"界面使用 FeatureCoverFlow 控件展示图片组合，因此需要创建一个该组合的 Item 界面，Item 界面主要显示一张图片，界面效果如图 4-3 所示。

【任务实施】

（1）创建"相册"界面 Item。在 res/layout 文件夹中创建布局文件 item_album.xml。

（2）放置界面控件。在布局文件中，放置一个 ImageView 控件用于显示相册中的图片，具体代码如文件 4-2 所示。

图 4-3 "相册"界面 Item

【文件 4-2】 item_album.xml

```
1   <?xml version="1.0" encoding="utf-8"?>
2   <FrameLayout xmlns:android="http://schemas.android.com/apk/res/android"
3       android:layout_width="@dimen/album_width"
4       android:layout_height="@dimen/album_height"
5       android:clickable="true"
6       android:foreground="@drawable/album_selector">
7       <ImageView
8           android:id="@+id/iv_img"
9           android:layout_width="match_parent"
10          android:layout_height="match_parent"
11          android:scaleType="centerCrop" />
12  </FrameLayout>
```

(3) 创建背景选择器。"相册"界面中的每张图片在被按下与弹起时，图片背景会有明显的区别，这种效果可以通过背景选择器实现。首先选中 drawable 文件夹，右击选择 New →Drawable resource file 选项，创建一个背景选择器 album_selector.xml，根据图片被按下与弹起时的状态切换它的背景颜色，当图片被按下时显示灰色背景（颜色值是#96000000），当图片弹起时显示透明背景，具体代码如文件 4-3 所示。

【文件 4-3】 album_selector.xml

```
1   <?xml version="1.0" encoding="utf-8"?>
2   <selector xmlns:android="http://schemas.android.com/apk/res/android">
3       <item android:state_pressed="true">
4           <color android:color="#96000000" />
5       </item>
6       <item>
7           <color android:color="@android:color/transparent" />
8       </item>
9   </selector>
```

(4) 创建 item_title.xml 文件。由于在"相册"界面的布局中放置了一个 TextSwitcher 控件显示图片的标题，该控件在实现 ViewFactory 接口中的 makeView()方法时必须返回一个 TextView 控件，因此需要在 res/layout 文件夹中创建一个 item_title.xml 文件放置返回的 TextView 控件，具体代码如文件 4-4 所示。

【文件 4-4】 item_title.xml

```
1   <?xml version="1.0" encoding="utf-8"?>
2   <TextView xmlns:android="http://schemas.android.com/apk/res/android"
3       android:layout_width="match_parent"
4       android:layout_height="wrap_content"
5       android:gravity="center"
6       android:textAppearance="?android:textAppearanceLargeInverse"
7       android:textColor="@android:color/white" />
```

【任务 4-3】 创建 AlbumBean

【任务分析】

由于"相册"界面主要显示一组图片,每张图片都有图片 Id 和标题 Id 属性,因此可以创建一个 AlbumBean 类存放图片的这些属性。

【任务实施】

在 com.itheima.album 包中创建一个 AlbumBean 类。在该类中创建图片属性,具体代码如文件 4-5 所示。

【文件 4-5】 AlbumBean.java

```
1  package com.itheima.album;
2  public class AlbumBean {
3      public int imgResId;            //图片 Id
4      public int titleResId;          //图片标题 Id
5      public AlbumBean(int imgResId, int titleResId) {
6          this.imgResId=imgResId;
7          this.titleResId=titleResId;
8      }
9  }
```

【任务 4-4】 "相册"界面 Adapter

【任务分析】

"相册"界面中显示的一组图片是通过 FeatureCoverFlow 控件实现的,因此需要创建一个数据适配器 AlbumAdapter 对该控件进行数据适配。

【任务实施】

(1) 创建 AlbumAdapter 类。在 com.itheima.album 包中创建一个 AlbumAdapter 类继承 BaseAdapter 类,并重写 getCount()、getItem()、getItemId()、getView()方法。

(2) 创建 ViewHolder 类。在 AlbumAdapter 类中创建一个 ViewHolder 类的 Item 界面上的图片控件对应的字段。

(3) 设置 Item 界面数据。在 getView()方法中,通过 inflate()方法加载"相册"界面的 Item 布局文件 item_album.xml,然后将图片设置到界面控件上,具体代码如文件 4-6 所示。

【文件 4-6】 AlbumAdapter.java

```
1  package com.itheima.album;
2  import android.content.Context;
3  import android.view.LayoutInflater;
4  import android.view.View;
5  import android.view.ViewGroup;
6  import android.widget.BaseAdapter;
```

```java
7    import android.widget.ImageView;
8    import java.util.ArrayList;
9    public class AlbumAdapter extends BaseAdapter {
10       private ArrayList<AlbumBean>dataList=new ArrayList<>();
11       private Context mContext;
12       public AlbumAdapter(Context context) {
13           mContext=context;
14       }
15       public void setData(ArrayList<AlbumBean>dataList) {
16           this.dataList=dataList;
17       }
18       @Override
19       public int getCount() {
20           return dataList.size();
21       }
22       @Override
23       public Object getItem(int position) {
24           return dataList.get(position);
25       }
26       @Override
27       public long getItemId(int position) {
28           return position;
29       }
30       @Override
31       public View getView(int position, View convertView, ViewGroup parent) {
32           if(convertView==null) {
33               LayoutInflater inflater=(LayoutInflater) mContext.
34                       getSystemService(Context.LAYOUT_INFLATER_SERVICE);
35               convertView=inflater.inflate(R.layout.item_album, null);
36               ViewHolder viewHolder=new ViewHolder();
37               viewHolder.iv_img=(ImageView) convertView.findViewById(R.id.iv_img);
38               convertView.setTag(viewHolder);
39           }
40           ViewHolder holder=(ViewHolder) convertView.getTag();
41           holder.iv_img.setImageResource(dataList.get(position).imgResId);
42           return convertView;
43       }
44       public class ViewHolder {
45           public ImageView iv_img;
46       }
47   }
```

【任务 4-5】 "相册"界面逻辑代码

【任务分析】

"相册"界面主要显示一组图片,滑动界面上的图片可以切换不同的图片到界面的中间位置,同时在图片下方会显示该图片的标题,点击界面上的任意一张图片会弹出对应图片的标题信息。

【任务实施】

（1）设置界面数据。在 AlbumActivity 中创建一个 initData()方法，在该方法中创建界面需要的图片数据。

（2）获取界面控件。在 AlbumActivity 中创建一个 init()方法，在该方法中获取"相册"界面的控件并设置相册控件的点击事件与滑动事件。

（3）双击"后退"键退出程序。在 AlbumActivity 中重写 onKeyDown()方法，在该方法中判断两次点击"后退"键的时间间隔是否小于 2s，如果小于 2s，则退出该程序；如果大于 2s，则提示"再按一次退出 3D 相册"。具体代码如文件 4-7 所示。

【文件 4-7】 AlbumActivity.java

```
1   package com.itheima.album;
2   import android.os.Bundle;
3   import android.support.v7.app.AppCompatActivity;
4   import android.view.KeyEvent;
5   import android.view.LayoutInflater;
6   import android.view.View;
7   import android.view.animation.Animation;
8   import android.view.animation.AnimationUtils;
9   import android.widget.AdapterView;
10  import android.widget.TextSwitcher;
11  import android.widget.TextView;
12  import android.widget.Toast;
13  import android.widget.ViewSwitcher;
14  import com.itheima.coverflow.ui.FeatureCoverFlow;
15  import java.util.ArrayList;
16  public class AlbumActivity extends AppCompatActivity {
17      private FeatureCoverFlow coverFlow;
18      private AlbumAdapter adapter;
19      private ArrayList<AlbumBean> dataList;
20      private TextSwitcher mTitle;
21      @Override
22      protected void onCreate(Bundle savedInstanceState) {
23          super.onCreate(savedInstanceState);
24          setContentView(R.layout.activity_album);
25          initData();
26          initView();
27      }
28      /**
29       * 初始化界面控件
30       */
31      private void initView() {
32          mTitle=(TextSwitcher) findViewById(R.id.ts_title);
33          mTitle.setFactory(new ViewSwitcher.ViewFactory() {
34              @Override
35              public View makeView() {
36                  LayoutInflater inflater=LayoutInflater.from(AlbumActivity.this);
37                  TextView title=(TextView) inflater.inflate(R.layout.item_title, null);
```

```
38              return title;
39          }
40      });
41      Animation in=AnimationUtils.loadAnimation(this, R.anim.slide_in_top);
42      Animation out=AnimationUtils.loadAnimation(this, R.anim.slide_out_bottom);
43      mTitle.setInAnimation(in);
44      mTitle.setOutAnimation(out);
45      coverFlow= (FeatureCoverFlow) findViewById(R.id.fcf_coverflow);
46      adapter=new AlbumAdapter(this);
47      adapter.setData(dataList);
48      coverFlow.setAdapter(adapter);
49      coverFlow.setOnItemClickListener(new AdapterView.OnItemClickListener() {
50          @Override
51          public void onItemClick(AdapterView<?>parent, View view, int position, long
52      id) {
53              if(position<dataList.size()) {
54                  Toast.makeText(AlbumActivity.this,
55                      getResources().getString(dataList.get(position).titleResId),
56                      Toast.LENGTH_SHORT).show();
57              }
58          }
59      });
60      coverFlow.setOnScrollPositionListener(new FeatureCoverFlow.
61      OnScrollPositionListener() {
62          @Override
63          public void onScrolledToPosition(int position) {
64              mTitle.setText(getResources().getString(dataList.get(position).
65                                                          titleResId));
66          }
67          @Override
68          public void onScrolling() {
69              mTitle.setText("");
70          }
71      });
72  }
73  /**
74   * 初始化界面数据
75   */
76  private void initData() {
77      dataList=new ArrayList<>();
78      dataList.add(new AlbumBean(R.drawable.img_1, R.string.title1));
79      dataList.add(new AlbumBean(R.drawable.img_2, R.string.title2));
80      dataList.add(new AlbumBean(R.drawable.img_3, R.string.title3));
81      dataList.add(new AlbumBean(R.drawable.img_4, R.string.title4));
82      dataList.add(new AlbumBean(R.drawable.img_5, R.string.title5));
83  }
84  protected long exitTime;      //记录第一次点击时的时间
85  @Override
86  public boolean onKeyDown(int keyCode, KeyEvent event) {
87      if(keyCode==KeyEvent.KEYCODE_BACK
```

```
88                    && event.getAction()==KeyEvent.ACTION_DOWN) {
89                if((System.currentTimeMillis() - exitTime)>2000) {
90                    Toast.makeText(AlbumActivity.this,"再按一次退出 3D 相册",
91                                                     Toast.LENGTH_SHORT).show();
92                    exitTime=System.currentTimeMillis();
93                } else {
94                    AlbumActivity.this.finish();
95                    System.exit(0);
96                }
97                return true;
98            }
99            return super.onKeyDown(keyCode, event);
100       }
101   }
```

（4）设置标题进出动画。在显示图片标题时，标题信息会从下向上缓缓进入界面，当标题消失时，标题信息会从上向下缓缓离开界面，因此首先需要在 res 文件夹中创建一个 anim 文件夹，然后在该文件夹中分别创建 slide_in_top.xml 文件与 slide_out_bottom.xml 文件以实现标题信息显示与消失对应的动画效果，具体代码如文件 4-8 与文件 4-9。

【文件 4-8】　slide_in_top.xml

```
1    <?xml version="1.0" encoding="utf-8"?>
2    <set xmlns:android="http://schemas.android.com/apk/res/android">
3        <translate
4            android:duration="@android:integer/config_mediumAnimTime"
5            android:fromYDelta="100%p"
6            android:toYDelta="0" />
7        <alpha
8            android:duration="@android:integer/config_mediumAnimTime"
9            android:fromAlpha="0.0"
10           android:toAlpha="1.0" />
11   </set>
```

【文件 4-9】　slide_out_bottom.xml

```
1    <?xml version="1.0" encoding="utf-8"?>
2    <set xmlns:android="http://schemas.android.com/apk/res/android">
3        <translate
4            android:duration="@android:integer/config_mediumAnimTime"
5            android:fromYDelta="0"
6            android:toYDelta="100%p" />
7        <alpha
8            android:duration="@android:integer/config_mediumAnimTime"
9            android:fromAlpha="1.0"
10           android:toAlpha="0.0" />
11   </set>
```

（5）修改 strings.xml 文件。由于"相册"界面的图片组合中一共有 5 张图片，每张图片都有对应的标题，因此需要在 res/values 文件夹中的 strings.xml 文件中创建 5 张图片对应的标题,具体代码如下所示。

```
<string name="title1">Girl</string>
<string name="title2">Spring Scenery</string>
<string name="title3">Summer Scenery</string>
<string name="title4">Autumn Scenery</string>
<string name="title5">Winter Scenery</string>
```

（6）修改清单文件。由于本项目有自己的图标,因此需要设置清单文件中的<application>标签的 icon 属性为@mipmap/ album_icon。

由于项目创建时默认都带有的标题栏不够美观,因此需要设置清单文件中的<application>标签中的 theme 属性为@style/Theme. AppCompat. Light. NoActionBar,删除标题栏。

由于"相册"界面在处于横屏时相对较清晰,因此需要设置 AlbumActivity 对应的<activity>标签中的 screenOrientation 属性为 landscape,固定该界面为横屏。

4.2　本章小结

本章主要讲解了 3D 效果的相册的实现过程,在这个实现过程中主要引入一个 library 库,调用该库中的自定义控件 FeatureCoverFlow 显示界面上的相册,同时该控件也可以使图片水平滑动,然后通过 TextSwitcher 控件显示图片的标题,最终实现整个 3D 效果的相册。在相册效果的实现过程中,代码量相对较少,逻辑简单,读者按照步骤完成即可。

【思考题】

1. 如何设置 3D 相册界面中每张图片的标题?
2. 如何把图片设置到 3D 相册界面上?

第 5 章 项 目 综 述

学习目标
- 了解头条项目的功能与模块结构
- 了解头条项目的界面交互效果

头条项目的数据源于"传智播客.黑马程序员"官网,该项目主要用于展示官网的热点新闻、各学科开班情况、技术视频以及个人信息等。本项目集合诸多热点技术,如下拉刷新、HelloCharts 图表库、BoomMenu 圆形菜单、BubbleViews 心形泡泡库、第三方视频播放等。本章将针对头条项目的整体功能进行简单介绍,第 6 章便可以开始真正的项目之旅了。

5.1 项目分析

5.1.1 项目名称

思政材料 5

黑马头条项目,简称"头条项目"。

5.1.2 项目概述

头条项目是一个新闻阅读类项目,其中包含新闻信息、各校区开班情况、就业薪资情况以及一些免费的技术视频等。同时,该项目还提供了一个"我"模块,主要用于展示用户信息以及一些娱乐功能,如日历、星座、涂鸦、地图等,供有兴趣的人进行研究。

5.1.3 开发环境

操作系统:
- Windows 系统

开发工具:
- JDK 8
- Android Studio 2.2.2+模拟器
- Tomcat 7.0.56

数据库:
- SQLite

API 版本：
- Android API 25

5.1.4 模块说明

头条项目主要分为四大功能模块，分别为首页模块、统计模块、视频模块、"我"模块，项目结构如图 5-1 所示。

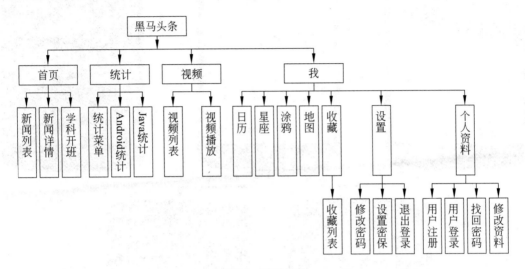

图 5-1 项目结构

从图 5-1 中可以看出，首页模块包含新闻列表、新闻详情、学科开班等功能，统计模块包含统计菜单、Android 统计、Java 统计等功能，视频模块包含视频列表与视频播放等功能，"我"模块包含日历、星座、涂鸦、地图、收藏、设置、个人资料七个功能，其中设置功能又包含修改密码、设置密保、退出登录三个功能，个人资料功能包含用户注册、用户登录、找回密码、修改资料四个功能。

5.2 效果展示

5.2.1 欢迎界面与主界面

程序成功启动后，首先会在欢迎界面停留几秒然后进入主界面，点击主界面底部的导航栏可以切换到统计界面、视频界面、"我"界面，如图 5-2 所示。

5.2.2 "新闻详情"界面与"Python 学科"界面

点击首页界面中的某个新闻条目或滑动广告栏中的某一广告图片会进入"新闻详情"界面，展示当前的新闻信息。当点击首页界面中的四个学科中的任意一个学科时，会跳转到对应学科的界面，展示该学科在全国各地的开班情况，如图 5-3 所示。

图 5-2 欢迎界面与主界面

5.2.3 "统计详情"界面

"统计"界面通过屏幕右下角的一个圆形菜单展示 9 个学科,这 9 个学科分别是 Android 学科、Java 学科、PHP 学科、Python 学科、C/C++ 学科、iOS 学科、前端与移动开发学科、UI 设计学科、网络营销学科。点击右下角的圆形菜单后出现 9 个学科的菜单。点击每个菜单会跳转到相应学科的"统计详情"界面,在此以 Android 学科和 Java 学科为例,分别展示 Android 学科与 Java 学科的薪资情况,效果如图 5-4 所示。

第 5 章 项目综述

图 5-3 "新闻详情"界面与"Python 学科"界面

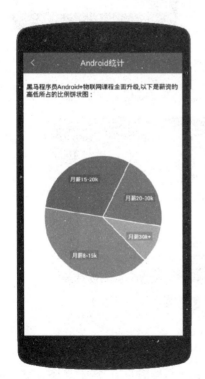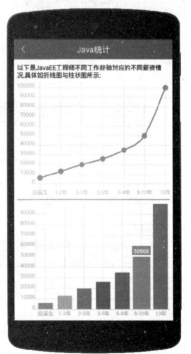

图 5-4 "统计详情"界面

5.2.4 "视频详情"界面

"视频"界面主要展示一些学科的视频信息,点击视频列表的某一条目会跳转到"视频详情"界面,"视频详情"界面主要由视频介绍、视频目录以及视频播放组成,"视频详情"界面效果如图 5-5 所示。

图 5-5 "视频详情"界面(1)

当点击"视频播放"界面的全屏按钮时,屏幕会变成横屏,效果如图 5-6 所示。

图 5-6 "视频详情"界面(2)

5.2.5 "我"界面

"我"界面主要展示一些个人信息(头像和账号),同时也有日历、星座、涂鸦、地图、收藏、设置等功能。登录成功时,点击"我"界面中的头像会跳转到"个人资料"界面,在未登录时,点击"我"界面中的头像会跳转到"登录"界面。登录成功时,"我"界面与未登录时的"我"界面,效果如图5-7所示。

图5-7 "我"界面

(1)点击"我"界面中的"日历""涂鸦""地图"等按钮时,会分别跳转到"日历"界面、"涂鸦"界面、"地图"界面,这些界面的效果如图5-8所示。

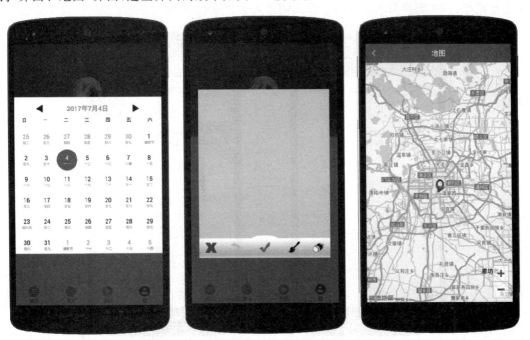

图5-8 "日历"界面、"涂鸦"界面与"地图"界面

(2)点击"我"界面中的"星座"按钮会跳转到"星座"界面,点击"星座"界面右上角的"切换"按钮会跳转到显示选择星座的界面,界面效果如图5-9所示。

图5-9 "星座"界面

(3)当用户未登录时,点击"我"界面中的收藏条目会提示"您还未登录,请先登录"。当用户登录时,点击"我"界面中的收藏条目会跳转到"收藏"界面。如果侧滑每个收藏的条目,则会出现红色的"删除"按钮,"收藏"界面的效果如图5-10所示。

图5-10 "收藏"界面

(4) 当用户未登录时,点击"我"界面中的设置条目会提示"您还未登录,请先登录"。当用户登录时,点击"我"界面中的设置条目会跳转到"设置"界面,"设置"界面效果如图 5-11 所示。

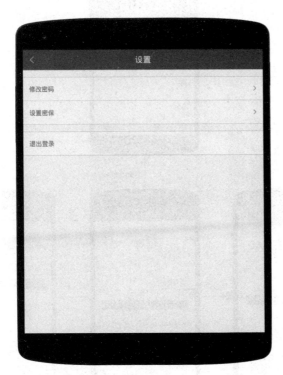

图 5-11 "设置"界面

5.2.6 "个人资料"界面

"个人资料"界面中包含登录、注册、找回密码、修改密码、设置密保、个人资料、修改昵称和签名、修改头像、修改性别等功能。下面分别介绍如何跳转到这些界面以及这些界面的效果图。

(1) 当用户未登录时,点击"我"界面中的头像会跳转到"登录"界面,如果没有登录账号则可以点击"快速注册"按钮进行注册,如果已有登录账号则输入正确的用户名和密码后即可登录,若忘记密码则可以点击"找回密码"按钮将密码找回,界面效果如图 5-12 所示。

(2) 点击"设置"界面中的"修改密码"与"设置密保"条目会分别跳转到"修改密码"界面与"设置密保"界面,"修改密码"界面与"设置密保"界面的效果如图 5-13 所示。

(3) 当用户登录时,点击"我"界面中的头像会跳转到"个人资料"界面,当用户点击"个人资料"界面的"头像""昵称""性别""签名"条目时,会分别跳转或弹出修改这些属性的界面,效果如图 5-14 所示。

图 5-12 "登录""注册""找回密码"界面

图 5-13 "修改密码"界面与"设置密保"界面

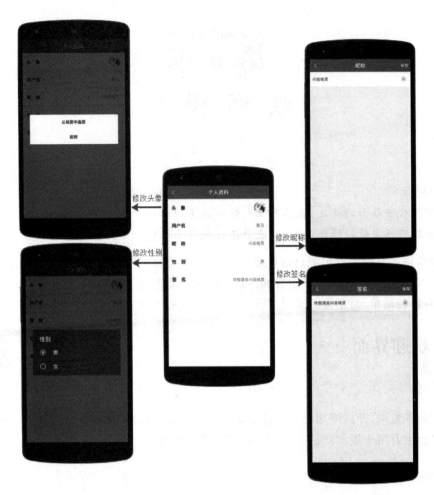

图 5-14 "个人资料"界面

5.3 本章小结

本章整体介绍了头条项目的功能、模块以及项目的界面效果,对这些,读者只需要在头脑中有一个简单的了解即可,后续章节会一一实现这些功能模块和界面的设计。

【思考题】

1. 头条项目有几个主要模块?
2. 在头条项目中如何进入登录界面与注册界面?

第 6 章 欢迎模块

学习目标

- 掌握欢迎界面的开发,能够独立制作欢迎界面
- 掌握底部导航栏模块的开发,能够搭建底部导航栏

欢迎模块主要用于展示项目的欢迎界面,由于主界面的功能较多,为了减轻主界面的实现压力,在欢迎模块中还实现了主界面的标题栏和底部导航栏。本章是头条项目的第一个模块,相对比较简单,对于读者来说只需轻装上阵、轻松对待。

6.1 欢迎界面

思政材料 6

任务综述

在实际开发中,开启应用程序时首先会呈现一个欢迎界面,本应用也不例外。头条项目的欢迎界面主要用于展示产品 LOGO 和公司的新闻信息。

【任务 6-1】 欢迎界面

【任务分析】

头条项目的欢迎界面效果如图 6-1 所示。

【任务实施】

(1) 创建项目。首先创建一个工程,并将其命名为 TopLine,指定包名为 com.itheima.topline。

(2) 导入界面图片。在 Android Studio 中切换到 Project 选项卡,在 res 文件夹中创建一个 drawable-hdpi 文件夹,将欢迎界面所需要的背景图片 launch_bg.png 导入 drawable-hdpi 文件夹,将项目的 icon 图标 app_icon.png 导入 mipmap-hdpi 文件夹。

(3) 创建欢迎界面。在程序中选中 com.itheima.topline 包,在该包下创建一个 activity 包,然后在 activity 包中创建一个 Empty Activity 类,命名为 SplashActivity,并将布局文件名指定为 activity_splash,具体代码如文

图 6-1 欢迎界面

件 6-1 所示。

【文件 6-1】 activity_splash.xml

```xml
1  <?xml version="1.0" encoding="utf-8"?>
2  <RelativeLayout xmlns:android="http://schemas.android.com/apk/res/android"
3      android:layout_width="match_parent"
4      android:layout_height="match_parent"
5      android:background="@drawable/launch_bg">
6  </RelativeLayout>
```

（4）修改 styles.xml 文件。项目创建时所有界面默认带有的蓝色标题栏不够美观，因此还需要在 res/values 文件夹的 styles.xml 文件中添加一个无标题栏的样式，具体代码如下：

```xml
<style name="AppTheme.NoActionBar">
    <item name="windowActionBar">false</item>
    <item name="windowNoTitle">true</item>
</style>
```

（5）修改清单文件。由于头条项目需要设置自己的图标与无标题栏的样式，因此需要修改清单文件中＜application＞标签中的 icon 与 theme 属性，具体代码如下：

```xml
android:icon="@mipmap/app_icon"
android:theme="@style/AppTheme.NoActionBar"
```

注意：由于 SplashActivity 为程序的启动界面，因此需要在清单文件中设置启动界面对应的 Activity 为 SplashActivity。

【任务 6-2】 欢迎界面逻辑代码

【任务分析】

欢迎界面主要展示产品 LOGO 与新闻信息，通常会在该界面停留一段时间后自动跳转到其他界面，因此需要在逻辑代码中设置欢迎界面暂停几秒(3s)后再跳转。

【任务实施】

在 SplashActivity 中创建 init() 方法，在该方法中使用 Timer 和 TimerTask 类设置欢迎界面延迟 3s 再跳转到主界面(MainActivity 所对应的界面,此界面目前为空白页面)，具体代码如文件 6-2 所示。

【文件 6-2】 SplashActivity.java

```java
1  package com.itheima.topline.activity;
2  public class SplashActivity extends AppCompatActivity {
3      @Override
4      protected void onCreate(Bundle savedInstanceState) {
5          super.onCreate(savedInstanceState);
```

```
6          setContentView(R.layout.activity_splash);
7          init();
8       }
9       private void init() {
10         //利用Timer让此界面延迟3s后再跳转,Timer中有一个线程,该线程不断执行task
11         Timer timer=new Timer();
12         //timertask实现runnable接口,TimerTask类表示一个在指定时间内执行的task
13         TimerTask task=new TimerTask() {
14             @Override
15             public void run() {
16                 Intent intent=new Intent(SplashActivity.this, MainActivity.class);
17                 startActivity(intent);
18                 SplashActivity.this.finish();
19             }
20         };
21         timer.schedule(task, 3000);          //设置该task在延迟3s后自动执行
22      }
23  }
```

6.2 导航栏

任务综述

根据设计图可知,头条项目中有一个底部导航栏,底部导航栏主要用于滑动切换不同的界面或者点击底部按钮时切换不同的界面,使用户操作起来比较方便快捷。

【任务6-3】 标题栏

【任务分析】

在头条项目中,大部分界面都有一个"后退"键和一个标题。为了便于代码重复利用,可以将"后退"键和标题抽取出来并单独放在一个布局文件(main_title_bar.xml)中,界面效果如图6-2所示。

图6-2 标题栏界面

【任务实施】

(1) 创建标题栏界面。在res/layout文件夹中创建一个布局文件main_title_bar.xml。在该布局文件中,放置2个TextView控件,分别用于显示后退键(后退键的样式采用背景选择器的方式)和当前界面标题(界面标题暂未设置,需要在代码中动态设置),并设置标题栏背景透明,具体代码如文件6-3所示。

【文件 6-3】 main_title_bar.xml

```xml
1   <?xml version="1.0" encoding="utf-8"?>
2   <RelativeLayout xmlns:android="http://schemas.android.com/apk/res/android"
3       android:id="@+id/title_bar"
4       android:layout_width="match_parent"
5       android:layout_height="45dp"
6       android:background="@android:color/transparent">
7       <TextView
8           android:id="@+id/tv_back"
9           android:layout_width="45dp"
10          android:layout_height="45dp"
11          android:layout_alignParentLeft="true"
12          android:layout_centerVertical="true"
13          android:background="@drawable/go_back_selector"
14          android:visibility="gone" />
15      <TextView
16          android:id="@+id/tv_main_title"
17          android:layout_width="wrap_content"
18          android:layout_height="wrap_content"
19          android:layout_centerInParent="true"
20          android:textColor="@android:color/white"
21          android:textSize="18sp" />
22  </RelativeLayout>
```

（2）创建背景选择器。标题栏界面中的返回键在按下与弹起时会有明显的区别，这种效果可以通过背景选择器实现。首先将图片 iv_back_selected.png 和 iv_back.png 导入 drawable-hdpi 文件夹，然后选中 drawable 文件夹，右击选择 New→Drawable resource file 选项，创建一个背景选择器 go_back_selector.xml，根据按钮按下和弹起的状态切换它的背景图片，给用户带来动态效果。当按钮按下时显示灰色图片（iv_back_selected.png），当按钮弹起时显示白色图片（iv_back.png），具体代码如文件 6-4 所示。

【文件 6-4】 go_back_selector.xml

```xml
1   <?xml version="1.0" encoding="utf-8"?>
2   <selector xmlns:android="http://schemas.android.com/apk/res/android">
3       <item android:drawable="@drawable/iv_back_selected"
4           android:state_pressed="true"/>
5       <item android:drawable="@drawable/iv_back"/>
6   </selector>
```

【任务 6-4】 底部导航栏

【任务分析】

根据前面介绍的设计图可知，此项目包含一个底部导航栏（即底部 4 个按钮），为了方便后续布局的搭建，创建一个底部导航栏 UI 的框架，界面效果如图 6-3 所示。

图 6-3　底部导航栏界面

【任务实施】

（1）导入界面图片。将底部导航栏所需的图片 home_normal.png、home_press.png、count_normal.png、count_press.png、video_normal.png、video_press.png、me_normal.png、me_press.png 导入 drawable-hdpi 文件夹。

（2）放置界面控件。在 activity_main.xml 文件中，通过<include>标签将 main_title_bar.xml（标题栏）引入，放置 4 个 RadioButton 控件，用于显示底部按钮；一个 ViewPager 控件，用于在加载每个底部按钮对应的界面后可以左右滑动，具体代码如文件 6-5 所示。

【文件 6-5】　activity_main.xml

```
1    <LinearLayout xmlns:android="http://schemas.android.com/apk/res/android"
2        android:layout_width="match_parent"
3        android:layout_height="match_parent"
4        android:orientation="vertical">
5        <include layout="@layout/main_title_bar"/>
6        <RelativeLayout
7            android:layout_width="match_parent"
8            android:layout_height="match_parent">
9            <RadioGroup
10               android:id="@+id/radioGroup"
11               android:layout_width="match_parent"
12               android:layout_height="50dp"
13               android:layout_alignParentBottom="true"
14               android:background="@android:color/white"
15               android:orientation="horizontal">
```

```
16          <RadioButton
17              android:id="@+id/rb_home"
18              style="@style/style_RadioButton"
19              android:checked="true"
20              android:drawableTop="@drawable/rb_home_selector"
21              android:text="首页"
22              android:textColor="@drawable/rb_text_selector" />
23          <RadioButton
24              android:id="@+id/rb_count"
25              style="@style/style_RadioButton"
26              android:drawableTop="@drawable/rb_count_selector"
27              android:text="统计"
28              android:textColor="@drawable/rb_text_selector" />
29          <RadioButton
30              android:id="@+id/rb_video"
31              style="@style/style_RadioButton"
32              android:drawableTop="@drawable/rb_video_selector"
33              android:text="视频"
34              android:textColor="@drawable/rb_text_selector" />
35          <RadioButton
36              android:id="@+id/rb_me"
37              style="@style/style_RadioButton"
38              android:drawableTop="@drawable/rb_me_selector"
39              android:text="我"
40              android:textColor="@drawable/rb_text_selector" />
41      </RadioGroup>
42      <android.support.v4.view.ViewPager
43          android:id="@+id/viewPager"
44          android:layout_width="match_parent"
45          android:layout_height="match_parent"
46          android:layout_above="@id/radioGroup" />
47  </RelativeLayout>
48 </LinearLayout>
```

（3）创建背景选择器。底部导航栏界面中的底部按钮在被选中与未被选中时会有明显的区别，这种效果可以通过背景选择器实现。右击选择 New→Drawable resource file 选项，创建一个背景选择器 rb_home_selector.xml，根据按钮被选中和未被选中的状态切换它的背景图片，给用户带来动态效果。当按钮未被选中时显示灰色图片（home_normal.png），当按钮被选中时显示红色图片（home_press.png），以此类推创建其余 3 个背景选择器 rb_count_selector.xml、rb_video_selector.xml、rb_me_selector.xml。由于这 4 个背景选择器的内容类似，因此只展示 rb_home_selector.xml 文件中的内容，具体代码如文件 6-6 所示。

【文件 6-6】 rb_home_selector.xml

```
1  <?xml version="1.0" encoding="utf-8"?>
2  <selector xmlns:android="http://schemas.android.com/apk/res/android">
3      <item android:state_checked="true" android:drawable="@drawable/home_press"/>
4      <item android:state_checked="false" android:drawable="@drawable/home_normal"/>
5  </selector>
```

（4）创建文字颜色选择器。底部导航栏界面中的底部按钮所对应的文字颜色在按钮被选中与未被选中时也会有变化，因此同样需要创建一个背景选择器 rb_text_selector.xml，用于控制文字颜色，当按钮未被选中时文字颜色为灰色（rdTextColorNormal），当按钮被选中时文字颜色为红色（rdTextColorPress），具体代码如文件 6-7 所示。

【文件 6-7】 rb_text_selector.xml

```xml
1  <?xml version="1.0" encoding="utf-8"?>
2  <selector xmlns:android="http://schemas.android.com/apk/res/android">
3      <item android:state_checked="true" android:color="@color/rdTextColorPress"/>
4      <item android:state_checked="false" android:color="@color/rdTextColorNormal"/>
5  </selector>
```

上述文件中的 rdTextColorPress、rdTextColorNormal 两个颜色值写在 res/values 文件夹的 colors.xml 文件中，便于以后多次调用，具体代码如下所示：

```xml
<color name="rdTextColorNormal">#999999</color>
<color name="rdTextColorPress">#eb413d</color>
```

（5）创建底部按钮的样式。由于底部导航栏的 4 个按钮具有相同的外观，因此需要为这 4 个按钮创建一个相同的样式，用于设置按钮的宽度、高度、比重、位置以及文字的大小。在项目的 res/values 文件夹的 styles.xml 文件中添加如下代码：

```xml
1  <style name="style_RadioButton">
2      <item name="android:layout_width">match_parent</item>
3      <item name="android:layout_height">wrap_content</item>
4      <item name="android:button">@null</item>
5      <item name="android:background">@null</item>
6      <item name="android:layout_weight">1</item>
7      <item name="android:gravity">center</item>
8      <item name="android:layout_gravity">center</item>
9      <item name="android:textSize">12sp</item>
10 </style>
```

【任务 6-5】 底部导航栏逻辑代码

【任务分析】

在底部导航栏中点击不同的按钮或者滑动 ViewPager 控件加载的页面会切换到不同的界面，因此需要为 4 个按钮设置选中状态的监听与 ViewPager 页面切换的监听。

【任务实施】

（1）获取界面控件。在 MainActivity 类中创建界面的初始化方法 initView()，用于获取标题栏、页面中间部分及底部导航栏界面所要使用到的控件，并设置 RadioGroup 选中状态的监听与 ViewPager 页面切换的监听。

（2）退出本应用。为了减少运行内存的占用，在不使用该应用时应该及时退出，因此需要在 MainActivity 中重写 onKeyDown() 方法，在该方法中调用 System.exit(0) 退出本应用，具体代码如文件 6-8 所示。

【文件 6-8】 MainActivity.java

```
1   package com.itheima.topline;
2   public class MainActivity extends AppCompatActivity {
3       private ViewPager viewPager;
4       private RadioGroup radioGroup;
5       @RequiresApi(api=Build.VERSION_CODES.HONEYCOMB)
6       private TextView tv_main_title;
7       private RelativeLayout rl_title_bar;
8       @Override
9       protected void onCreate(Bundle savedInstanceState) {
10          super.onCreate(savedInstanceState);
11          setContentView(R.layout.activity_main);
12          initView();
13      }
14      private void initView() {
15          tv_main_title=(TextView) findViewById(R.id.tv_main_title);
16          tv_main_title.setText("首页");
17          rl_title_bar=(RelativeLayout) findViewById(R.id.title_bar);
18          rl_title_bar.setBackgroundColor(getResources().getColor(R.color.
19                                                      rdTextColorPress));
20          radioGroup=(RadioGroup) findViewById(R.id.radioGroup);
21          //RadioGroup 选中状态改变监听
22          radioGroup.setOnCheckedChangeListener(new RadioGroup.
23          OnCheckedChangeListener() {
24              @Override
25              public void onCheckedChanged(RadioGroup group, int checkedId) {
26                  switch(checkedId) {
27                      case R.id.rb_home:
28                          //setCurrentItem()方法中第二个参数控制页面切换动画,true:打开,false:关闭
29                          viewPager.setCurrentItem(0, false);
30                          break;
31                      case R.id.rb_count:
32                          viewPager.setCurrentItem(1, false);
33                          break;
34                      case R.id.rb_video:
35                          viewPager.setCurrentItem(2, false);
36                          break;
37                      case R.id.rb_me:
38                          viewPager.setCurrentItem(3, false);
39                          break;
40                  }
41              }
42          });
43          viewPager=(ViewPager) findViewById(R.id.viewPager);
44          //ViewPager 页面切换监听
45          viewPager.setOnPageChangeListener(new ViewPager.OnPageChangeListener()
```

```java
46      {
47          @Override
48          public void onPageScrolled(int position, float positionOffset,
49                                      int positionOffsetPixels) {
50          }
51          @Override
52          public void onPageSelected(int position) {
53              switch(position) {
54                  case 0:
55                      radioGroup.check(R.id.rb_home);
56                      tv_main_title.setText("首页");
57                      rl_title_bar.setVisibility(View.VISIBLE);
58                      break;
59                  case 1:
60                      radioGroup.check(R.id.rb_count);
61                      tv_main_title.setText("统计");
62                      rl_title_bar.setVisibility(View.VISIBLE);
63                      break;
64                  case 2:
65                      radioGroup.check(R.id.rb_video);
66                      tv_main_title.setText("视频");
67                      rl_title_bar.setVisibility(View.VISIBLE);
68                      break;
69                  case 3:
70                      radioGroup.check(R.id.rb_me);
71                      rl_title_bar.setVisibility(View.GONE);
72                      break;
73              }
74          }
75          @Override
76          public void onPageScrollStateChanged(int state) {
77          }
78      });
79  }
80  protected long exitTime;        //记录第一次点击时的时间
81  @Override
82  public boolean onKeyDown(int keyCode, KeyEvent event) {
83      if(keyCode==KeyEvent.KEYCODE_BACK
84              && event.getAction()==KeyEvent.ACTION_DOWN) {
85          if((System.currentTimeMillis()-exitTime)>2000) {
86              Toast.makeText(MainActivity.this, "再按一次退出黑马头条",
87                      Toast.LENGTH_SHORT).show();
88              exitTime=System.currentTimeMillis();
89          } else {
90              MainActivity.this.finish();
91              System.exit(0);
92          }
93          return true;
94      }
95      return super.onKeyDown(keyCode, event);
96  }
97 }
```

6.3 本章小结

本章主要讲解了头条项目的欢迎模块,欢迎模块主要包括欢迎界面与底部导航栏,这两个界面相对比较简单,读者按照步骤完成即可。

【思考题】

1. 如何实现欢迎界面暂停几秒后再跳转?
2. 如何搭建底部导航栏?

第 7 章 首页模块

学习目标

- 掌握服务器的搭建,能够独立制作服务器
- 掌握工具类的开发,能够快速创建工具类
- 掌握首页模块的开发,能够独立制作首页模块
- 掌握新闻详情模块的开发,实现新闻详情界面的显示
- 掌握 Python 学科模块的开发,实现 Python 学科界面的显示

首页模块主要用于展示水平滑动广告栏、新闻列表以及四个学科(Python、Java、PHP、Android)按钮。点击广告栏或者新闻列表时会跳转到新闻详情界面,点击"学科"按钮会跳转到对应学科并展示该学科的开班计划。本章将针对首页模块进行详细讲解。

7.1 搭建服务器

任务综述

思政材料 7

由于头条项目所使用的后台数据量比较大,因此该项目以 Tomcat 7.0.56 为例搭建一个小型的简易服务器。在该服务器的 ROOT 文件夹中存放首页数据信息与图片信息。

【任务 7-1】 首页广告栏数据

【任务分析】

首页广告栏由广告图片与广告数据组成,其中图片是通过在 Tomcat 的 ROOT 文件夹中创建一个图片文件夹 banner 进行存放的,数据是通过在 ROOT 文件夹中创建一个 home_ad_list_data.json 文件进行存放的。

【任务实施】

(1) 创建存放首页广告栏图片的文件夹。在 Tomcat 的 ROOT 文件夹中创建一个 topline 文件夹,用于存放头条项目中的所有数据,在 topline 文件夹中创建一个 img 文件夹,在 img 文件夹中创建一个 banner 文件夹,用于存放首页广告栏图片。

(2) 创建首页广告栏数据。在 ROOT/topline 文件夹中创建一个 home_ad_list_data.json 文件,用于存放首页广告栏数据,具体代码如文件 7-1 所示。

【文件 7-1】 home_ad_list_data.json

```
1  [
2    {
3      "id":1,
4      "type":3,
5      "newsName":"黑马超级惠狼季",
6      "newsTypeName":"",
7      "img1":"http://172.16.43.62:8080/topline/img/banner/banner1.png",
8      "img2":"",
9      "img3":"",
10     "newsUrl":"http://www.itheima.com/special/hyjzly/index.shtml?tj"
11   },
12   {
13     "id":2,
14     "type":3,
15     "newsName":"网络营销",
16     "newsTypeName":"",
17     "img1":"http://172.16.43.62:8080/topline/img/banner/banner2.png",
18     "img2":"",
19     "img3":"",
20     "newsUrl":"http://www.itheima.com/special/mhmzly/"
21   },
22   {
23     "id":3,
24     "type":3,
25     "newsName":"UI设计",
26     "newsTypeName":"",
27     "img1":"http://172.16.43.62:8080/topline/img/banner/banner3.png",
28     "img2":"",
29     "img3":"",
30     "newsUrl":"http://www.itheima.com/special/szicd/"
31   },
32   {
33     "id":4,
34     "type":3,
35     "newsName":"人工智能+Python",
36     "newsTypeName":"",
37     "img1":"http://172.16.43.62:8080/topline/img/banner/banner4.png",
38     "img2":"",
39     "img3":"",
40     "picUrl":"http://www.itheima.com/special/pythonzly/"
41   }
42 ]
```

注意：上述文件中的 IP 地址 172.16.43.62 为本机的 IP 地址，在使用时需要将其修改为自己计算机的 IP 地址，否则无法访问数据。此处的 IP 地址不可设置为 localhost。

【任务 7-2】 首页新闻列表数据

【任务分析】

首页新闻列表由新闻图片与新闻数据组成,其中图片是通过在 Tomcat 的 ROOT 文件夹中创建一个图片文件夹 homenews 进行存放的,数据是通过在 ROOT 文件夹中创建一个 home_news_list_data.json 文件进行存放的。

【任务实施】

(1)创建存放首页新闻列表图片的文件夹。在 Tomcat 的 ROOT/topline/img 文件夹中创建一个 homenews 文件夹,用于存放首页新闻列表图片。

(2)创建首页新闻列表数据。在 ROOT/topline 文件夹中创建一个 home_news_list_data.json 文件,用于存放首页新闻列表数据,具体代码如文件 7-2 所示。

【文件 7-2】 home_news_list_data.json

```
1   [
2     {
3       "id":1,
4       "type":1,
5       "newsName":"黑马程序员成功入驻合肥,开启江淮学子高薪时代",
6       "newsTypeName":"黑马新闻",
7       "img1":"http://172.16.43.62:8080/topline/img/homenews/home_news1.png",
8       "img2":"",
9       "img3":"",
10      "newsUrl":"http://bbs.itheima.com/thread-343752-1-1.html?tj"
11    },
12    {
13      "id":2,
14      "type":2,
15      "newsName":"iOS 课程全新升级",
16      "newsTypeName":"黑马推荐",
17      "img1":"http://172.16.43.62:8080/topline/img/homenews/home_recom_ios1.png",
18      "img2":"http://172.16.43.62:8080/topline/img/homenews/home_recom_ios2.png",
19      "img3":"http://172.16.43.62:8080/topline/img/homenews/home_recom_ios3.png",
20      "newsUrl":"http://www.itheima.com/special/hmiospro/?tj"
21    },
22    {
23      "id":3,
24      "type":1,
25      "newsName":"在线学院(来在线学院 不离职、不离校)",
26      "newsTypeName":"黑马新闻",
27      "img1":"http://172.16.43.62:8080/topline/img/homenews/home_news_college.png",
28      "img2":"",
29      "img3":"",
30      "newsUrl":"http://www.itheima.com/special/online/index.html?tj"
31    },
32    {
```

```
33      "id":4,
34      "type":2,
35      "newsName":"人工智能+Python将革新更多领域",
36      "newsTypeName":"黑马推荐",
37      "img1":"http://172.16.43.62:8080/topline/img/homenews/home_recom_python1.png",
38      "img2":"http://172.16.43.62:8080/topline/img/homenews/home_recom_python2.png",
39      "img3":"http://172.16.43.62:8080/topline/img/homenews/home_recom_python3.png",
40      "newsUrl":"http://www.itheima.com/special/pythonzly/?tj"
41    },
42    {
43      "id":5,
44      "type":1,
45      "newsName":"全栈工程师顺势而生,打造精通前端+后端+移动端的精英型全栈工程师",
46      "newsTypeName":"黑马新闻",
47      "img1":"http://172.16.43.62:8080/topline/img/homenews/home_news_q.png",
48      "img2":"",
49      "img3":"",
50      "newsUrl":"http://www.itheima.com/special/stackzly/?zxdt"
51    }
52 ]
```

7.2 工具类

任务综述

由于头条项目中会遇到多个界面重复使用同一个方法的情况,因此为了减少重复代码量,需要创建一个工具类存放这些方法,当任意一个界面需要调用该方法时,直接调用工具类中的方法即可。

【任务 7-3】 创建 Constant 类

【任务分析】

由于头条项目是从 Tomcat(一台小型服务器)中获取数据并展现在各个界面上的,因此需要创建一个 Constant 类存放各个界面向服务器请求数据所需要的接口地址。

【任务实施】

选中 com.itheima.topline 包,在该包下创建 utils 包。在 utils 包中创建一个 Constant 类。在该类中创建首页广告栏与新闻列表接口地址,具体代码如文件 7-3 所示。

【文件 7-3】 Constant.java

```
1  package com.itheima.topline.utils;
2  public class Constant {
3      //内网接口
4      public static final String WEB_SITE="http://172.16.43.62:8080/topline";
5      //首页滑动广告接口
```

```
6       public static final String REQUEST_AD_URL="/home_ad_list_data.json";
7       //首页新闻列表接口
8       public static final String REQUEST_NEWS_URL="/home_news_list_data.json";
9   }
```

【任务 7-4】 创建 JsonParse 类

【任务分析】

由于头条项目的 Tomcat 服务器使用的是 JSON 数据，因此需要创建一个 JsonParse 类解析从服务器中获取的 JSON 数据。

【任务实施】

（1）创建 JsonParse 类。在 com.itheima.topline.utils 包中创建一个 JsonParse 类。

（2）添加 gson 库。在 Android Studio 中，选中项目后右击选择 Open Module Settings 选项，然后选择 Dependencies 选项卡，单击右上角的绿色加号选择 Library dependency 选项，把 com.google.code.gson:gson:2.2.4 库加入主项目。

（3）调用 gson 库解析 JSON 数据。JsonParse 类主要用于解析从服务器获取的 JSON 数据，具体代码如文件 7-4 所示。

【文件 7-4】 JsonParse.java

```
1   package com.itheima.topline.utils;
2   public class JsonParse {
3       private static JsonParse instance;
4       private JsonParse() {
5       }
6       public static JsonParse getInstance() {
7           if(instance==null) {
8               instance=new JsonParse();
9           }
10          return instance;
11      }
12  }
```

【任务 7-5】 创建 UtilsHelper 类

【任务分析】

由于头条项目的部分界面会用到相同的功能，因此为了减少项目中的重复代码量与便于后续调用，需要把实现这些功能的方法抽取出来并放在一个工具类 UtilsHelper 中。

【任务实施】

（1）创建 UtilsHelper 类。在 com.itheima.topline.utils 包中创建一个 UtilsHelper 类。

（2）获取屏幕宽度。在 UtilsHelper 类中，创建一个 getScreenWidth()方法用于获取屏幕宽度，在首页界面设置广告栏宽度时调用，具体代码如文件 7-5 所示。

【文件 7-5】 UtilsHelper.java

```java
package com.itheima.topline.utils;
public class UtilsHelper {
    /**
     * 获得屏幕宽度
     */
    public static int getScreenWidth(Context context) {
        WindowManager wm= (WindowManager) context.getSystemService(Context.WINDOW_SERVICE);
        DisplayMetrics outMetrics=new DisplayMetrics();
        wm.getDefaultDisplay().getMetrics(outMetrics);
        return outMetrics.widthPixels;
    }
}
```

7.3 首页

任务综述

在头条项目的开发中，程序在经过欢迎界面后会直接进入主界面，也就是首页界面。首页界面分为上下两部分，上部分通过 ViewPager 与 Fragment 实现滑动广告展示，下部分通过一个自定义的 WrapRecyclerView 控件展示新闻推荐信息。由于头条项目使用的是 Tomcat 搭建的一台小型服务器，因此首页界面的所有数据必须存放在 Tomcat 根目录的 JSON 文件中，并通过解析 JSON 文件获取数据填充界面。

【任务 7-6】 水平滑动广告栏界面

【任务分析】

水平滑动广告栏主要用于展示广告信息或者活动信息，由 ViewPager 控件、TextView 控件以及一个自定义的线性布局 ViewPagerIndicator 组成，界面效果如图 7-1 所示。

图 7-1　水平滑动广告栏界面

【任务实施】

(1) 创建水平滑动广告栏界面。在 res/layout 文件夹中创建一个布局文件 main_adbanner.xml。

(2) 放置界面控件。在布局文件中,放置一个 ViewPager 控件用于显示左右滑动的广告图片,由于广告栏左下角的标题与右下角的小圆点都随着图片的滑动而发生变化,因此需要一个 TextView 控件与一个自定义的 ViewPagerIndicator 控件分别显示标题和小圆点,具体代码如文件 7-6 所示。

【文件 7-6】 main_adbanner.xml

```
1   <?xml version="1.0" encoding="utf-8"?>
2   <RelativeLayout xmlns:android="http://schemas.android.com/apk/res/android"
3       android:id="@+id/adbanner_layout"
4       android:layout_width="match_parent"
5       android:layout_height="wrap_content">
6       <android.support.v4.view.ViewPager
7           android:id="@+id/slidingAdvertBanner"
8           android:layout_width="fill_parent"
9           android:layout_height="fill_parent"
10          android:layout_alignParentLeft="true"
11          android:layout_alignParentTop="true"
12          android:layout_marginBottom="1dp"
13          android:background="@android:color/black"
14          android:gravity="center" />
15      <LinearLayout
16          android:layout_width="fill_parent"
17          android:layout_height="wrap_content"
18          android:layout_alignParentBottom="true"
19          android:background="#82000000">
20          <TextView
21              android:id="@+id/tv_advert_title"
22              android:layout_width="0dp"
23              android:layout_height="wrap_content"
24              android:layout_marginLeft="8dp"
25              android:layout_marginRight="8dp"
26              android:layout_weight="1"
27              android:ellipsize="end"
28              android:gravity="left|center_vertical"
29              android:padding="4dp"
30              android:singleLine="true"
31              android:textColor="@android:color/white"
32              android:textSize="14sp" />
33          <com.itheima.topline.view.ViewPagerIndicator
34              android:id="@+id/advert_indicator"
35              android:layout_width="wrap_content"
36              android:layout_height="fill_parent"
37              android:layout_gravity="center_vertical"
38              android:layout_marginRight="@dimen/activity_horizontal_margin"
```

```
39                    android:gravity="right|center_vertical"
40                    android:padding="4dp"/>
41          </LinearLayout>
42  </RelativeLayout>
```

在上述文件中,第 6~14 行代码用于添加 ViewPager 控件。需要注意的是,在布局文件中添加 ViewPager 控件需要写出 ViewPager 的全路径。

(3) 自定义 ViewPagerIndicator 控件。在实际开发中,很多时候 Android 自带的控件都不能满足用户的需求,此时就需要自定义一个控件。在头条项目中,水平滑动广告栏底部的小圆点控件就需要通过自定义控件实现,因此需要在程序中选中 com.itheima.topline 包,在该包下创建一个 view 包,然后在 view 包中创建一个 ViewPagerIndicator 类并继承 LinearLayout 类,具体代码如文件 7-7 所示。

【文件 7-7】 ViewPagerIndicator.java

```
1   package com.itheima.topline.view;
2   public class ViewPagerIndicator extends LinearLayout {
3       private int mCount;                    //小圆点的个数
4       private int mIndex;                    //当前小圆点的位置
5       private Context context;
6       public ViewPagerIndicator(Context context) {
7           this(context, null);
8       }
9       public ViewPagerIndicator(Context context, AttributeSet attrs) {
10          super(context, attrs);
11          this.context=context;
12      }
13      /**
14       * 设置滑动到当前小圆点时其他圆点的位置
15       */
16      public void setCurrentPostion(int currentIndex) {
17          mIndex=currentIndex;              //当前小圆点
18          this.removeAllViews();            //移除界面上存在的 view
19          int pex=context.getResources().getDimensionPixelSize(
20                                  R.dimen.view_indicator_padding);
21          for(int i=0; i<this.mCount; i++) {
22              //创建一个 ImageView 控件放置小圆点
23              ImageView imageView=new ImageView(context);
24              if(mIndex==i) {               //滑动到的当前界面
25                  //设置小圆点的图片为白色图片
26                  imageView.setImageResource(R.drawable.indicator_on);
27              }else {
28                  //设置小圆点的图片为灰色图片
29                  imageView.setImageResource(R.drawable.indicator_off);
30              }
31              imageView.setPadding(pex, 0, pex, 0);     //设置小圆点图片上下左右的 padding
32              this.addView(imageView);      //把小圆点添加到自定义 ViewPagerIndicator 控件上
33          }
```

```
34        }
35    /**
36     * 设置小圆点的数目
37     */
38    public void setCount(int count) {
39        this.mCount=count;
40    }
41 }
```

（4）修改 dimens.xml 文件。由于在 ViewPagerIndicator 类中使用到 view_indicator_padding 设置界面上圆点之间的距离，因此需要在 res/values 文件夹的 dimens.xml 文件中添加如下代码：

```
<dimen name="view_indicator_padding">5dp</dimen>
```

（5）创建 indicator_on.xml 和 indicator_off.xml 文件。在自定义控件 ViewPagerIndicator 中分别有一个白色和一个灰色的小圆点图片，这两张图片是通过在 drawable 文件夹下分别创建 indicator_on.xml 和 indicator_off.xml 两个文件来实现的，具体代码如文件 7-8 和文件 7-9 所示。

【文件 7-8】 indicator_on.xml

```
1  <?xml version="1.0" encoding="utf-8"?>
2  <shape xmlns:android="http://schemas.android.com/apk/res/android"
3      android:shape="oval">
4      <size android:height="6dp" android:width="6dp"/>
5      <solid android:color="#E9E9E9"/>
6  </shape>
```

【文件 7-9】 indicator_off.xml

```
1  <?xml version="1.0" encoding="utf-8"?>
2  <shape xmlns:android="http://schemas.android.com/apk/res/android"
3      android:shape="oval">
4      <size android:height="6dp" android:width="6dp"/>
5      <solid android:color="#BCBCBC"/>
6  </shape>
```

在上述代码中，shape 用于设定形状，可以用于选择器和布局，shape 默认为矩形（rectangle），可设置为椭圆形（oval）、线性形状（line）、环形（ring）；size 表示大小，可设置宽和高；solid 表示内部填充色。

【任务 7-7】 首页界面

【任务分析】

首页界面主要由水平滑动广告栏、四个学科按钮以及一个新闻列表组成，广告栏主要用于展示广告或活动信息，四个学科按钮分别展示 Python 学科、Java 学科、PHP 学科、

Android 学科。新闻列表主要用于展示新闻信息,界面效果如图 7-2 所示。

【任务实施】

(1) 创建首页界面。由于首页界面分为两部分,一部分是滑动广告栏与学科按钮,另一部分是新闻列表,因此需要在 res/layout 文件夹中分别创建两个布局文件 fragment_home.xml 与 head_view.xml,在 head_view.xml 文件中,通过<include>标签将 main_adbanner.xml(广告栏)引入。

(2) 导入界面图片。将首页界面所需的图片 python_icon.png、java_icon.png、php_icon.png、android_icon.png 导入 drawable-hdpi 文件夹。

(3) 引入第三方下拉刷新。在实际开发中,很多时候都需要展示一些比较炫酷的功能的效果,如果在程序中直接开发,则代码量会大幅增加,也会耗费大量的开发时间,因此开发人员通常会引入第三方框架以实现这些功能,头条项目中的下拉刷新功能就是通过引入第三方下拉刷新框架实现的。

在 Android Studio 中,选择 File→New→Import Module 选项把下拉刷新的框架导入项目,选中项目后右击选择 Open Module Settings 选项后选择 Dependencies 选项卡,单击右上角的绿色加号并选择 Module Dependency 选项,把下拉刷新框架加入主项目,下拉刷新框架的详情如图 7-3 所示。

图 7-2 首页界面

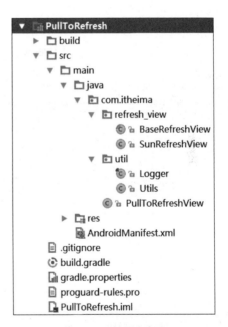

图 7-3 下拉刷新框架

(4) 在 fragment_home.xml 文件中放置界面控件。在 fragment_home.xml 文件中,放置一个自定义的 PullToRefreshView 控件,用于显示下拉刷新;一个自定义的 WrapRecyclerView 控件,用于显示新闻列表信息,具体代码如文件 7-10 所示。

【文件 7-10】 fragment_home.xml

```xml
1  <?xml version="1.0" encoding="utf-8"?>
2  <LinearLayout xmlns:android="http://schemas.android.com/apk/res/android"
3      android:layout_width="match_parent"
4      android:layout_height="match_parent"
5      android:orientation="vertical">
6      <com.itheima.PullToRefreshView
7          android:id="@+id/pull_to_refresh"
8          android:layout_width="match_parent"
9          android:layout_height="match_parent"
10         android:background="#f6f6f6">
11         <com.itheima.topline.view.WrapRecyclerView
12             android:id="@+id/recycler_view"
13             android:layout_width="match_parent"
14             android:layout_height="match_parent"
15             android:divider="@null"
16             android:dividerHeight="0dp"
17             android:fadingEdge="none" />
18     </com.itheima.PullToRefreshView>
19 </LinearLayout>
```

（5）在 head_view.xml 文件中放置界面控件。在 head_view.xml 文件中，放置 4 个 ImageView 控件，用于显示 4 个学科所对应的图标，放置 4 个 TextView 控件分别显示 4 个学科所对应的文字，具体代码如文件 7-11 所示。

【文件 7-11】 head_view.xml

```xml
1  <?xml version="1.0" encoding="utf-8"?>
2  <LinearLayout xmlns:android="http://schemas.android.com/apk/res/android"
3      android:layout_width="match_parent"
4      android:layout_height="wrap_content"
5      android:background="@android:color/white"
6      android:orientation="vertical">
7      <include layout="@layout/main_adbanner" />
8      <LinearLayout
9          android:layout_width="fill_parent"
10         android:layout_height="wrap_content"
11         android:layout_marginBottom="8dp"
12         android:layout_marginTop="8dp"
13         android:orientation="horizontal">
14         <LinearLayout
15             android:id="@+id/ll_python"
16             android:layout_width="0dp"
17             android:layout_height="wrap_content"
18             android:layout_weight="1"
19             android:gravity="center"
20             android:orientation="vertical">
21             <ImageView
22                 android:layout_width="40dp"
23                 android:layout_height="40dp"
```

```xml
24              android:background="@drawable/python_icon" />
25          <TextView
26              android:layout_width="wrap_content"
27              android:layout_height="wrap_content"
28              android:layout_marginTop="4dp"
29              android:text="Python学科"
30              android:textSize="14sp" />
31      </LinearLayout>
32      <LinearLayout
33          android:id="@+id/ll_java"
34          android:layout_width="0dp"
35          android:layout_height="wrap_content"
36          android:layout_weight="1"
37          android:gravity="center"
38          android:orientation="vertical">
39          <ImageView
40              android:layout_width="40dp"
41              android:layout_height="40dp"
42              android:background="@drawable/java_icon" />
43          <TextView
44              android:layout_width="wrap_content"
45              android:layout_height="wrap_content"
46              android:layout_marginTop="4dp"
47              android:text="Java学科"
48              android:textSize="14sp" />
49      </LinearLayout>
50      <LinearLayout
51          android:id="@+id/ll_php"
52          android:layout_width="0dp"
53          android:layout_height="wrap_content"
54          android:layout_weight="1"
55          android:gravity="center"
56          android:orientation="vertical">
57          <ImageView
58              android:layout_width="40dp"
59              android:layout_height="40dp"
60              android:background="@drawable/php_icon" />
61          <TextView
62              android:layout_width="wrap_content"
63              android:layout_height="wrap_content"
64              android:layout_marginTop="4dp"
65              android:text="PHP学科"
66              android:textSize="14sp" />
67      </LinearLayout>
68      <LinearLayout
69          android:id="@+id/ll_android"
70          android:layout_width="0dp"
71          android:layout_height="wrap_content"
72          android:layout_weight="1"
73          android:gravity="center"
```

```
74              android:orientation="vertical">
75              <ImageView
76                  android:layout_width="40dp"
77                  android:layout_height="40dp"
78                  android:background="@drawable/android_icon" />
79              <TextView
80                  android:layout_width="wrap_content"
81                  android:layout_height="wrap_content"
82                  android:layout_marginTop="4dp"
83                  android:text="Android学科"
84                  android:textSize="14sp" />
85          </LinearLayout>
86      </LinearLayout>
87  </LinearLayout>
```

【任务 7-8】 自定义控件 WrapRecyclerView

【任务分析】

在头条项目中，首页分为两部分，一部分是由广告栏与学科按钮组成的界面（头部界面）；另一部分是一个新闻列表界面；若想把两部分结合起来，需要自定义一个 WrapRecyclerView 控件，用于展示新闻列表，然后调用自定义控件中添加头部界面的方法组成完整的首页界面。

【任务实施】

（1）添加 recyclerview-v7 库。由于新闻列表用到 recyclerview-v7 包中的 RecyclerView 类，因此需要在 Android Studio 中选中项目后右击选择 Open Module Settings 选项，然后选择 Dependencies 选项卡，单击右上角的绿色加号并选择 Library dependency 选项，然后找到 com.android.support:recyclerview-v7 库并添加到项目中。

（2）创建自定义控件 WrapRecyclerView。在 com.itheima.topline.view 包中创建一个 WrapRecyclerView 类并继承 RecyclerView 类，具体代码如文件 7-12 所示。

【文件 7-12】 WrapRecyclerView.java

```
1   package com.itheima.topline.view;
2   public class WrapRecyclerView extends RecyclerView {
3       private WrapAdapter mWrapAdapter;
4       private boolean shouldAdjustSpanSize;
5       //临时头部 View 集合,用于存储没有设置 Adapter 之前添加的头部
6       private ArrayList<View>mTmpHeaderView=new ArrayList<>();
7       public WrapRecyclerView(Context context) {
8           super(context);
9       }
10      public WrapRecyclerView(Context context, AttributeSet attrs) {
11          super(context, attrs);
12      }
13      public WrapRecyclerView(Context context, AttributeSet attrs, int defStyle) {
```

```java
14          super(context, attrs, defStyle);
15      }
16      @Override
17      public void setAdapter(Adapter adapter) {
18          if(adapter instanceof WrapAdapter) {
19              mWrapAdapter= (WrapAdapter) adapter;
20              super.setAdapter(adapter);
21          } else {
22              mWrapAdapter=new WrapAdapter(adapter);
23              for(View view : mTmpHeaderView) {
24                  mWrapAdapter.addHeaderView(view);
25              }
26              if(mTmpHeaderView.size()>0) {
27                  mTmpHeaderView.clear();
28              }
29              super.setAdapter(mWrapAdapter);
30          }
31          if(shouldAdjustSpanSize) {
32              mWrapAdapter.adjustSpanSize(this);
33          }
34          getWrappedAdapter().registerAdapterDataObserver(mDataObserver);
35          mDataObserver.onChanged();
36      }
37      @Override
38      public WrapAdapter getAdapter() {
39          return mWrapAdapter;
40      }
41      public Adapter getWrappedAdapter() {
42          if(mWrapAdapter==null) {
43              throw new IllegalStateException("You must set a adapter before!");
44          }
45          return mWrapAdapter.getWrappedAdapter();
46      }
47      public void addHeaderView(View view) {
48          if(null==view) {
49              throw new IllegalArgumentException("the view to add must not be null!");
50          } else if(mWrapAdapter==null) {
51              mTmpHeaderView.add(view);
52          } else {
53              mWrapAdapter.addHeaderView(view);
54          }
55      }
56      @Override
57      public void setLayoutManager(LayoutManager layout) {
58          super.setLayoutManager(layout);
59          if(layout instanceof GridLayoutManager || layout instanceof StaggeredGrid_
            LayoutManager){
60              this.shouldAdjustSpanSize=true;
61          }
62      }
```

```
63  private final AdapterDataObserver mDataObserver=new AdapterDataObserver() {
64      @Override
65      public void onChanged() {
66          if(mWrapAdapter !=null) {
67              mWrapAdapter.notifyDataSetChanged();
68          }
69      }
70      @Override
71      public void onItemRangeInserted(int positionStart, int itemCount) {
72          mWrapAdapter.notifyItemRangeInserted(positionStart, itemCount);
73      }
74      @Override
75      public void onItemRangeChanged(int positionStart, int itemCount) {
76          mWrapAdapter.notifyItemRangeChanged(positionStart, itemCount);
77      }
78      @Override
79      public void onItemRangeRemoved(int positionStart, int itemCount) {
80          mWrapAdapter.notifyItemRangeRemoved(positionStart, itemCount);
81      }
82      @Override
83      public void onItemRangeMoved(int fromPosition, int toPosition, int itemCount)
84      {
85          mWrapAdapter.notifyItemMoved(fromPosition, toPosition);
86      }
87  };
88  }
```

（3）创建 WrapAdapter 类。由于自定义的 WrapRecyclerView 控件需要对添加的头部界面进行设置，因此需要在程序中选中 com.itheima.topline 包，在该包下创建一个 adapter 包，然后在 adapter 包中创建一个 WrapAdapter 类继承 RecyclerView.Adapter＜RecyclerView.ViewHolder＞类，具体代码如文件 7-13 所示。

【文件 7-13】 WrapAdapter.java

```
1   package com.itheima.topline.adapter;
2   public class WrapAdapter<T extends RecyclerView.Adapter>extends
3   RecyclerView.Adapter<RecyclerView.ViewHolder>{
4       private final T mRealAdapter;
5       private boolean isStaggeredGrid;
6       private static final int BASE_HEADER_VIEW_TYPE=-1<<10;
7       private ArrayList<FixedViewInfo>mHeaderViewInfos=new ArrayList<>();
8       public class FixedViewInfo {
9           public View view;
10          public int viewType;
11      }
12      public WrapAdapter(T adapter) {
13          super();
14          mRealAdapter=adapter;
15      }
16      public T getWrappedAdapter() {
```

```java
17          return mRealAdapter;
18      }
19      public void addHeaderView(View view) {
20          if(null==view) {
21              throw new IllegalArgumentException("the view to add must not be null!");
22          }
23          final FixedViewInfo info=new FixedViewInfo();
24          info.view=view;
25          info.viewType=BASE_HEADER_VIEW_TYPE+mHeaderViewInfos.size();
26          mHeaderViewInfos.add(info);
27          notifyDataSetChanged();
28      }
29      public void adjustSpanSize(RecyclerView recycler) {
30          if(recycler.getLayoutManager() instanceof GridLayoutManager) {
31              final GridLayoutManager layoutManager= (GridLayoutManager)
32              recycler.getLayoutManager();
33              layoutManager.setSpanSizeLookup(new GridLayoutManager.SpanSizeLookup() {
34                  @Override
35                  public int getSpanSize(int position) {
36                      boolean isHeaderOrFooter=isHeaderPosition(position);
37                      return isHeaderOrFooter ? layoutManager.getSpanCount() : 1;
38                  }
39              });
40          }
41          if(recycler.getLayoutManager() instanceof StaggeredGridLayoutManager) {
42              this.isStaggeredGrid=true;
43          }
44      }
45      private boolean isHeader(int viewType) {
46          return viewType>=BASE_HEADER_VIEW_TYPE
47          && viewType< (BASE_HEADER_VIEW_TYPE+mHeaderViewInfos.size());
48      }
49      private boolean isHeaderPosition(int position) {
50          return position<mHeaderViewInfos.size();
51      }
52      @Override
53      public RecyclerView.ViewHolder onCreateViewHolder(ViewGroup viewGroup,
54      int viewType) {
55          if(isHeader(viewType)) {
56              int whichHeader=Math.abs(viewType - BASE_HEADER_VIEW_TYPE);
57              View headerView=mHeaderViewInfos.get(whichHeader).view;
58              return createHeaderFooterViewHolder(headerView);
59          } else {
60              return mRealAdapter.onCreateViewHolder(viewGroup, viewType);
61          }
62      }
63      private RecyclerView.ViewHolder createHeaderFooterViewHolder(View view) {
64          if(isStaggeredGrid) {
65              StaggeredGridLayoutManager.LayoutParams params=new
66                  StaggeredGridLayoutManager.LayoutParams(
```

```
67                    StaggeredGridLayoutManager.LayoutParams.MATCH_PARENT,
68                    StaggeredGridLayoutManager.LayoutParams.WRAP_CONTENT);
69            params.setFullSpan(true);
70            view.setLayoutParams(params);
71        }
72        return new RecyclerView.ViewHolder(view) {
73        };
74    }
75    @SuppressWarnings("unchecked")
76    @Override
77    public void onBindViewHolder(RecyclerView.ViewHolder viewHolder,
78            int position) {
79        if(position<mHeaderViewInfos.size()) {
80        } else if(position<mHeaderViewInfos.size()+mRealAdapter.
81                getItemCount()){
82            mRealAdapter.onBindViewHolder(viewHolder,
83                    position -mHeaderViewInfos.size());
84        }
85    }
86    @Override
87    public int getItemCount() {
88        return mHeaderViewInfos.size()+mRealAdapter.getItemCount();
89    }
90    @Override
91    public int getItemViewType(int position) {
92        if(isHeaderPosition(position)) {
93            return mHeaderViewInfos.get(position).viewType;
94        } else {
95            return mRealAdapter.getItemViewType(position -mHeaderViewInfos.size());
96        }
97    }
98 }
```

【任务 7-9】 首页界面 Item

【任务分析】

首页界面使用 WrapRecyclerView 控件展示新闻列表,因此需要创建一个该列表的 Item 界面。Item 分为两种形式,一种是新闻类型,界面上显示一个新闻标题、一张新闻图片以及一个新闻类型;另一种是推荐类型,界面上显示三张推荐信息图片、一个推荐信息标题以及一个推荐类型,界面效果如图 7-4 所示。

图 7-4 首页界面 Item

【任务实施】

(1) 创建首页界面 Item。在 res/layout 文件夹中,分别创建布局文件 home_item_one.xml 与 home_item_two.xml。

(2) 放置界面控件。在布局文件中,放置一个 ImageView 控件用于显示新闻图片;两个 TextView 控件分别用于显示新闻名称与新闻类型,具体代码如文件 7-14 所示。

【文件 7-14】 home_item_one.xml

```
1   <?xml version="1.0" encoding="utf-8"?>
2   <RelativeLayout xmlns:android="http://schemas.android.com/apk/res/android"
3       android:layout_width="match_parent"
4       android:layout_height="100dp"
5       android:layout_marginLeft="8dp"
6       android:layout_marginRight="8dp"
7       android:layout_marginTop="4dp"
8       android:background="@drawable/item_bg_selector"
9       android:padding="8dp">
10      <ImageView
11          android:id="@+id/iv_img"
12          android:layout_width="100dp"
13          android:layout_height="80dp"
14          android:layout_alignParentLeft="true"
15          android:layout_centerVertical="true"
16          android:scaleType="fitXY" />
17      <LinearLayout
18          android:layout_width="wrap_content"
19          android:layout_height="wrap_content"
20          android:layout_centerVertical="true"
21          android:layout_marginLeft="10dp"
22          android:layout_toRightOf="@id/iv_img"
23          android:orientation="vertical">
24          <TextView
25              android:id="@+id/tv_name"
26              android:layout_width="fill_parent"
27              android:layout_height="wrap_content"
28              android:textColor="@android:color/black"
29              android:textSize="14sp" />
30          <TextView
31              android:id="@+id/tv_newsType_name"
32              android:layout_width="fill_parent"
33              android:layout_height="wrap_content"
34              android:layout_marginTop="8dp"
35              android:textSize="12sp" />
36      </LinearLayout>
37  </RelativeLayout>
```

(3) 放置 home_item_two.xml 文件中的控件。在布局文件中,放置 3 个 ImageView 控件用于显示推荐信息的图片;2 个 TextView 控件分别用于显示新闻名称与新闻类型,具体代码如文件 7-15 所示。

【文件 7-15】 home_item_two.xml

```xml
1   <?xml version="1.0" encoding="utf-8"?>
2   <LinearLayout xmlns:android="http://schemas.android.com/apk/res/android"
3       android:layout_width="match_parent"
4       android:layout_height="wrap_content"
5       android:layout_marginLeft="8dp"
6       android:layout_marginRight="8dp"
7       android:layout_marginTop="4dp"
8       android:background="@drawable/item_bg_selector"
9       android:orientation="vertical"
10      android:padding="8dp">
11      <TextView
12          android:id="@+id/tv_name"
13          android:layout_width="fill_parent"
14          android:layout_height="wrap_content"
15          android:textColor="@android:color/black"
16          android:textSize="14sp" />
17      <LinearLayout
18          android:layout_width="fill_parent"
19          android:layout_height="80dp"
20          android:layout_gravity="center_vertical"
21          android:layout_marginTop="8dp"
22          android:orientation="horizontal">
23          <ImageView
24              android:id="@+id/iv_img1"
25              android:layout_width="0dp"
26              android:layout_height="fill_parent"
27              android:layout_weight="1"
28              android:scaleType="fitXY" />
29          <ImageView
30              android:id="@+id/iv_img2"
31              android:layout_width="0dp"
32              android:layout_height="fill_parent"
33              android:layout_marginLeft="6dp"
34              android:layout_weight="1"
35              android:scaleType="fitXY" />
36          <ImageView
37              android:id="@+id/iv_img3"
38              android:layout_width="0dp"
39              android:layout_height="fill_parent"
40              android:layout_marginLeft="6dp"
41              android:layout_weight="1"
42              android:scaleType="fitXY" />
43      </LinearLayout>
44      <TextView
45          android:id="@+id/tv_newsType_name"
46          android:layout_width="wrap_content"
```

```
47          android:layout_height="wrap_content"
48          android:layout_marginTop="8dp"
49          android:textSize="12sp" />
50  </LinearLayout>
```

（4）创建 Item 界面的背景选择器。Item 界面背景的四个角是椭圆形的,并且在按下与弹起时,背景颜色会有明显的区别,这种效果可以通过背景选择器实现。选中 drawable 文件夹,右击选择 New→Drawable resource file 选项,创建一个背景选择器 item_bg_selector.xml,根据按钮按下和弹起的状态变换它的背景颜色,给用户带来动态效果。当背景被按下时显示灰色(#fafafa),当背景弹起时显示白色(#ffffff),具体代码如文件 7-16 所示。

【文件 7-16】 item_bg_selector.xml

```
1   <?xml version="1.0" encoding="utf-8"?>
2   <selector xmlns:android="http://schemas.android.com/apk/res/android">
3       <item android:state_pressed="true">
4           <shape android:shape="rectangle">
5               <corners android:radius="5dp"/>
6               <solid android:color="#fafafa"></solid>
7           </shape>
8       </item>
9       <item android:state_pressed="false">
10          <shape android:shape="rectangle">
11              <corners android:radius="5dp"/>
12              <solid android:color="#ffffff"></solid>
13          </shape>
14      </item>
15  </selector>
```

【任务 7-10】 创建 NewsBean

【任务分析】

由于首页的新闻信息包含新闻 Id、新闻类型、新闻名称、新闻类型名称、跟帖数量、新闻图片 1、新闻图片 2、新闻图片 3、新闻链接等属性,同时首页的广告栏信息包含广告栏 Id、广告图片、广告标题、广告链接等属性,因此可以创建一个 NewsBean 类存放新闻信息和广告栏信息的属性。

【任务实施】

选中 com.itheima.topline 包,在该包下创建 bean 包。在 bean 包中创建一个 NewsBean 类并实现 Serializable 接口。在该类中创建新闻信息与广告栏信息的所有属性,具体代码如文件 7-17 所示。

【文件 7-17】 NewsBean.java

```
1   package com.itheima.topline.bean;
2   public class NewsBean implements Serializable {
```

```java
3      private static final long serialVersionUID=1L;
4      private int id;                          //新闻 Id
5      //若 type 为 1(黑马新闻)则显示一张图片的布局,若为 2(黑马推荐)则显示三张图片的布局
6      private int type;
7      private String newsName;                 //新闻名称
8      private String newsTypeName;             //新闻类型,是黑马新闻还是黑马推荐
9      private String img1;                     //新闻图片 1
10     private String img2;                     //新闻图片 2
11     private String img3;                     //新闻图片 3
12     private String newsUrl;                  //新闻链接
13     public int getType() {
14         return type;
15     }
16     public void setType(int type) {
17         this.type=type;
18     }
19     public String getNewsTypeName() {
20         return newsTypeName;
21     }
22     public void setNewsTypeName(String newsTypeName) {
23         this.newsTypeName=newsTypeName;
24     }
25     public int getId() {
26         return id;
27     }
28     public void setId(int id) {
29         this.id=id;
30     }
31     public String getNewsName() {
32         return newsName;
33     }
34     public void setNewsName(String newsName) {
35         this.newsName=newsName;
36     }
37     public String getImg1() {
38         return img1;
39     }
40     public void setImg1(String img1) {
41         this.img1=img1;
42     }
43     public String getImg2() {
44         return img2;
45     }
46     public void setImg2(String img2) {
47         this.img2=img2;
48     }
49     public String getImg3() {
50         return img3;
51     }
52     public void setImg3(String img3) {
```

```
53            this.img3=img3;
54        }
55        public String getNewsUrl() {
56            return newsUrl;
57        }
58        public void setNewsUrl(String newsUrl) {
59            this.newsUrl=newsUrl;
60        }
61    }
```

【任务 7-11】 创建 AdBannerFragment

【任务分析】

由于首页界面的广告栏用到了 ViewPager 控件，因此需要创建一个 AdBannerFragment 类设置 ViewPager 控件中的数据。

【任务实施】

（1）创建 AdBannerFragment 类。选中 com.itheima.topline 包，在该包下创建 fragment 包。在 fragment 包中创建一个 AdBannerFragment 类并继承 android.support.v4.app.Fragment 类（Android Studio 自带了一种创建 Fragment 的方法，在 Fragment 创建后会默认重写多个无用的方法，因此为了方便起见，直接通过继承类的方式创建一个 Fragment，重写所需方法）。

（2）添加图片加载框架 glide-3.7.0.jar。在 Project 选项卡下的 app 中有一个 libs 文件夹，如果没有则新建一个，然后把 glide-3.7.0.jar 包复制到 libs 文件夹中，选中 glide-3.7.0.jar 包，右击选择 Add As Library 选项，然后弹出一个对话框，把该 jar 包放在 app 的项目中即可。

（3）创建 AdBannerFragment 对应的视图。由于需要创建 AdBannerFragment 对应的视图，因此需要重写 onCreateView() 方法，在该方法中创建滑动广告栏的视图，具体代码如文件 7-18 所示。

【文件 7-18】 AdBannerFragment.java

```
1   package com.itheima.topline.fragment;
2   public class AdBannerFragment extends Fragment {
3       private NewsBean nb;                    //广告
4       private ImageView iv;                   //图片
5       public static AdBannerFragment newInstance(Bundle args) {
6           AdBannerFragment af=new AdBannerFragment();
7           af.setArguments(args);
8           return af;
9       }
10      @Override
11      public void onCreate(Bundle savedInstanceState) {
12          super.onCreate(savedInstanceState);
```

```
13          Bundle arg=getArguments();
14          nb=(NewsBean) arg.getSerializable("ad");       //获取一个新闻对象
15      }
16      @Override
17      public void onActivityCreated(Bundle savedInstanceState) {
18          super.onActivityCreated(savedInstanceState);
19      }
20      @Override
21      public void onResume() {
22          super.onResume();
23          if(nb !=null) {
24              //调用 Glide 框架加载图片
25              Glide
26                      .with(getActivity())
27                      .load(nb.getImg1())
28                      .error(R.mipmap.ic_launcher)
29                      .into(iv);
30          }
31      }
32      @Override
33      public View onCreateView(LayoutInflater inflater, ViewGroup container,
34                  Bundle savedInstanceState) {
35          iv=new ImageView(getActivity());
36          ViewGroup.LayoutParams lp=new ViewGroup.LayoutParams(
37                  ViewGroup.LayoutParams.FILL_PARENT,
38                  ViewGroup.LayoutParams.FILL_PARENT);
39          iv.setLayoutParams(lp);                        //设置图片的宽和高参数
40          iv.setScaleType(ImageView.ScaleType.FIT_XY);   //把图片填满整个控件
41          iv.setOnClickListener(new View.OnClickListener() {
42              @Override
43              public void onClick(View v) {
44              }
45          });
46          return iv;
47      }
48  }
```

【任务7-12】 创建 AdBannerAdapter

【任务分析】

由于广告栏用到了 ViewPager 控件，因此需要创建一个数据适配器 AdBannerAdapter 对 ViewPager 控件进行数据适配。

【任务实施】

（1）创建 AdBannerAdapter 类。在 com.itheima.topline.adapter 包中，创建一个 AdBannerAdapter 类继承 FragmentStatePagerAdapter 类并实现 OnTouchListener 接口。

（2）创建设置数据方法 setData()。在 AdBannerAdapter 类中创建一个 setData()方法，通过接收 List 集合设置界面数据，具体代码如文件 7-19 所示。

【文件 7-19】 AdBannerAdapter.java

```
1   package com.itheima.topline.adapter;
2   public class AdBannerAdapter extends FragmentStatePagerAdapter implements
3       View.OnTouchListener {
4       private List<NewsBean> abl;
5       public AdBannerAdapter(FragmentManager fm) {
6           super(fm);
7           abl=new ArrayList<NewsBean>();
8       }
9       /**
10       * 设置数据更新界面
11       */
12      public void setData(List<NewsBean> abl) {
13          this.abl=abl;
14          notifyDataSetChanged();
15      }
16      @Override
17      public Fragment getItem(int index) {
18          Bundle args=new Bundle();
19          if(abl.size()>0)
20              args.putSerializable("ad", abl.get(index % abl.size()));
21          return AdBannerFragment.newInstance(args);
22      }
23      @Override
24      public int getCount() {
25          return Integer.MAX_VALUE;
26      }
27      /**
28       * 返回数据集的真实容量大小
29       */
30      public int getSize() {
31          return abl==null ? 0 : abl.size();
32      }
33      /**
34       * 获取广告名称
35       */
36      public String getTitle(int index) {
37          return abl==null ? null : abl.get(index).getNewsName();
38      }
39      @Override
40      public int getItemPosition(Object object) {
41          //防止刷新结果显示列表时出现缓存数据,重载这个函数,使其默认返回 POSITION_NONE
42          return POSITION_NONE;
43      }
44      @Override
45      public boolean onTouch(View v, MotionEvent event) {
```

```
46            return false;
47        }
48 }
```

【任务 7-13】 首页界面 Adapter

【任务分析】

首页界面的新闻列表是通过 WrapRecyclerView 控件展示的，因此需要创建一个数据适配器 HomeListAdapter 对 WrapRecyclerView 控件进行数据适配。由于 Item 类型分为两种，因此需要在 HomeListAdapter 中根据新闻类型判断需要显示哪种类型的 Item。

【任务实施】

（1）创建 HomeListAdapter 类。在 com.itheima.topline.adapter 包中创建一个 HomeListAdapter 类继承 RecyclerView.Adapter＜RecyclerView.ViewHolder＞类，并重写 onCreateViewHolder()、onBindViewHolder()、getItemViewType()、getItemCount()方法。在 onCreateViewHolder()方法中根据新闻类型设置 XML 布局。

（2）创建 TypeOneViewHolder 类和 TypeTwoViewHolder 类。由于 Item 界面是根据新闻类型的不同而加载不同的布局文件的，因此需要在 HomeListAdapter 中分别创建 TypeOneViewHolder 类与 TypeTwoViewHolder 类用于获取两个界面上的控件，具体代码如文件 7-20 所示。

【文件 7-20】 HomeListAdapter.java

```
1  package com.itheima.topline.adapter;
2  public class HomeListAdapter extends RecyclerView.Adapter<RecyclerView.
3  ViewHolder>{
4      private List<NewsBean>newsList;
5      private static final int TYPE_ONE=1;         //一张图片的样式
6      private static final int TYPE_TWO=2;         //三张图片的样式
7      private Context context;
8      public HomeListAdapter(Context context) {
9          this.context=context;
10     }
11     public void setData(List<NewsBean>newsList) {
12         this.newsList=newsList;
13         notifyDataSetChanged();
14     }
15     @Override
16     public RecyclerView.ViewHolder onCreateViewHolder(ViewGroup viewGroup, int
17  viewType){
18         if(viewType==TYPE_TWO) {
19             View view=LayoutInflater.from(viewGroup.getContext()).inflate(
20                                 R.layout.home_item_two, viewGroup, false);
21             TypeTwoViewHolder viewHolder=new TypeTwoViewHolder(view);
22             return viewHolder;
```

```java
23        } else {
24            View view=LayoutInflater.from(viewGroup.getContext()).inflate(
25                            R.layout.home_item_one, viewGroup, false);
26            TypeOneViewHolder viewHolder=new TypeOneViewHolder(view);
27            return viewHolder;
28        }
29    }
30    @Override
31    public void onBindViewHolder(final RecyclerView.ViewHolder holder, int i) {
32        if(newsList==null) return;
33        final NewsBean bean=newsList.get(i);
34        if(holder instanceof TypeOneViewHolder) {
35            ((TypeOneViewHolder) holder).tv_name.setText(bean.getNewsName());
36            ((TypeOneViewHolder) holder).tv_news_type_name.setText(bean.getNewsTypeName());
37            Glide
38                    .with(context)
39                    .load(bean.getImg1())
40                    .error(R.mipmap.ic_launcher)
41                    .into(((TypeOneViewHolder) holder).iv_img);
42        } else if(holder instanceof TypeTwoViewHolder) {
43            ((TypeTwoViewHolder) holder).tv_name.setText(bean.getNewsName());
44            ((TypeTwoViewHolder) holder).tv_news_type_name.setText(
45            bean.getNewsTypeName());
46            Glide
47                    .with(context)
48                    .load(bean.getImg1())
49                    .error(R.mipmap.ic_launcher)
50                    .into(((TypeTwoViewHolder) holder).iv_img1);
51            Glide
52                    .with(context)
53                    .load(bean.getImg2())
54                    .error(R.mipmap.ic_launcher)
55                    .into(((TypeTwoViewHolder) holder).iv_img2);
56            Glide
57                    .with(context)
58                    .load(bean.getImg3())
59                    .error(R.mipmap.ic_launcher)
60                    .into(((TypeTwoViewHolder) holder).iv_img3);
61        }
62        holder.itemView.setOnClickListener(new View.OnClickListener() {
63            @Override
64            public void onClick(View view) {
65            }
66        });
67    }
68    @Override
69    public int getItemViewType(int position) {
70        if(1==newsList.get(position).getType()) {
71            return TYPE_ONE;           //一张图片的类型
72        } else if(2==newsList.get(position).getType()) {
```

```
73              return TYPE_TWO;            //三张图片的类型
74          } else {
75              return TYPE_ONE;
76          }
77      }
78      @Override
79      public int getItemCount() {
80          return newsList==null ? 0 : newsList.size();
81      }
82      public class TypeOneViewHolder extends RecyclerView.ViewHolder {
83          public TextView tv_name, tv_news_type_name;
84          public ImageView iv_img;
85          public TypeOneViewHolder(View itemView) {
86              super(itemView);
87              tv_name=(TextView) itemView.findViewById(R.id.tv_name);
88              tv_news_type_name=(TextView) itemView.findViewById(R.id.
99                          tv_newsType_name);
90              iv_img=(ImageView) itemView.findViewById(R.id.iv_img);
91          }
92      }
93      public class TypeTwoViewHolder extends RecyclerView.ViewHolder {
94          public TextView tv_name, tv_news_type_name;
95          public ImageView iv_img1, iv_img2, iv_img3;
96          public TypeTwoViewHolder(View itemView) {
97              super(itemView);
98              tv_name=(TextView) itemView.findViewById(R.id.tv_name);
99              tv_news_type_name= (TextView) itemView.findViewById(R.id.
100                         tv_newsType_name);
101             iv_img1= (ImageView) itemView.findViewById(R.id.iv_img1);
102             iv_img2= (ImageView) itemView.findViewById(R.id.iv_img2);
103             iv_img3= (ImageView) itemView.findViewById(R.id.iv_img3);
104         }
105     }
106 }
```

【任务7-14】 首页界面逻辑代码

【任务分析】

在首页界面中需要编写广告栏逻辑代码和新闻列表逻辑代码，由于广告栏每隔一段时间会自动切换到下一张图片，因此可以创建一个线程实现。广告栏与新闻列表的数据是从Tomcat服务器中获取，并通过JSON解析这些数据并加载到对应的界面。

【任务实施】

（1）创建HomeFragment类。在com.itheima.topline.fragment包中创建一个HomeFragment类。在该类中，创建界面控件的初始化方法initView()，在该方法中获取页面布局中需要用到的UI控件并初始化。

（2）添加 okhttp 库。由于在头条项目中需要使用 OkHttpClient 类向服务器请求数据，因此需要选中项目后右击选择 Open Module Settings 选项后再选择 Dependencies 选项卡，单击右上角的绿色加号并选择 Library dependency 选项，然后找到 com.squareup.okhttp：okhttp：2.0.0 库并添加到项目中。

（3）从服务器获取数据。在 HomeFragment 中分别创建 getADData() 和 getNewsData() 方法，用于从 Tomcat 服务器中获取广告栏与新闻的数据。

（4）广告栏自动滑动时间间隔。创建一个 AdAutoSlidThread 线程设置广告栏的滑动时间间隔，具体代码如文件 7-21 所示。

【文件 7-21】 HomeFragment.java

```java
1   package com.itheima.topline.fragment;
2   public class HomeFragment extends Fragment {
3       private PullToRefreshView mPullToRefreshView;
4       private WrapRecyclerView recyclerView;
5       public static final int REFRESH_DELAY=1000;
6       private ViewPager adPager;                    //广告
7       private ViewPagerIndicator vpi;               //小圆点
8       private TextView tvAdName;                    //广告名称
9       private View adBannerLay;                    //广告条容器
10      private AdBannerAdapter ada;                  //适配器
11      public static final int MSG_AD_SLID=1;        //广告自动滑动
12      public static final int MSG_AD_OK=2;          //获取广告数据
13      public static final int MSG_NEWS_OK=3;        //获取新闻数据
14      private MHandler mHandler;                    //事件捕获
15      private LinearLayout ll_python;
16      private OkHttpClient okHttpClient;
17      private HomeListAdapter adapter;
18      public HomeFragment() {
19      }
20      @Override
21      public View onCreateView(LayoutInflater inflater, ViewGroup container,
22                              Bundle savedInstanceState) {
23          okHttpClient=new OkHttpClient();
24          mHandler=new MHandler();
25          getADData();
26          getNewsData();
27          View view=initView(inflater, container);
28          return view;
29      }
30      private View initView(LayoutInflater inflater, ViewGroup container) {
31          View view=inflater.inflate(R.layout.fragment_home, container, false);
32          recyclerView= (WrapRecyclerView) view.findViewById(R.id.recycler_view);
33          recyclerView.setLayoutManager(new LinearLayoutManager(getContext()));
34          View headView=inflater.inflate(R.layout.head_view, container, false);
35          recyclerView.addHeaderView(headView);
36          adapter=new HomeListAdapter(getActivity());
37          recyclerView.setAdapter(adapter);
```

```java
38          mPullToRefreshView= (PullToRefreshView) view.findViewById(
39                          R.id.pull_to_refresh);
40          mPullToRefreshView.setOnRefreshListener(new PullToRefreshView.
41                                          OnRefreshListener() {
42              @Override
43              public void onRefresh() {
44                  mPullToRefreshView.postDelayed(new Runnable() {
45                      @Override
46                      public void run() {
47                          mPullToRefreshView.setRefreshing(false);
48                          getADData();
49                          getNewsData();
50                      }
51                  }, REFRESH_DELAY);
52              }
53          });
54          adBannerLay=headView.findViewById(R.id.adbanner_layout);
55          adPager= (ViewPager) headView.findViewById(R.id.slidingAdvertBanner);
56          vpi= (ViewPagerIndicator) headView.findViewById(R.id.advert_indicator);
57          tvAdName= (TextView) headView.findViewById(R.id.tv_advert_title);
58          ll_python= (LinearLayout) headView.findViewById(R.id.ll_python);
59          adPager.setLongClickable(false);
60          ada=new AdBannerAdapter(getActivity().getSupportFragmentManager());
61          adPager.setAdapter(ada);
62          adPager.setOnTouchListener(ada);
63          adPager.setOnPageChangeListener(new ViewPager.OnPageChangeListener() {
64              @Override
65              public void onPageSelected(int index) {
66                  if(ada.getSize()>0) {
67                      if(ada.getTitle(index % ada.getSize()) !=null) {
68                          tvAdName.setText(ada.getTitle(index % ada.getSize()));
69                      }
70                      vpi.setCurrentPostion(index % ada.getSize());
71                  }
72              }
73              @Override
74              public void onPageScrolled(int arg0, float arg1, int arg2) {
75              }
76              @Override
77              public void onPageScrollStateChanged(int arg0) {
78              }
79          });
80          resetSize();
81          setListener();
82          new AdAutoSlidThread().start();
83          return view;
84      }
85      private void setListener() {
```

```java
86              ll_python.setOnClickListener(new View.OnClickListener() {
87                  @Override
88                  public void onClick(View view) {
89                  }
90              });
91          }
92          /**
93           * 事件捕获
94           */
95          class MHandler extends Handler {
96              @Override
97              public void dispatchMessage(Message msg) {
98                  super.dispatchMessage(msg);
99                  switch(msg.what) {
100                     case MSG_AD_SLID:
101                         if(ada.getCount()>0) {
102                             adPager.setCurrentItem(adPager.getCurrentItem()+1);
103                         }
104                         break;
105                     case MSG_AD_OK:
106                         if(msg.obj !=null) {
107                             String adResult= (String) msg.obj;
108                             List<NewsBean>adl=JsonParse.getInstance().
109                                         getAdList(adResult);
110                             if(adl !=null) {
111                                 if(adl.size()>0) {
112                                     ada.setData(adl);
113                                     tvAdName.setText(adl.get(0).getNewsName());
114                                     vpi.setCount(adl.size());
115                                     vpi.setCurrentPostion(0);
116                                 }
117                             }
118                         }
119                         break;
120                     case MSG_NEWS_OK:
121                         if(msg.obj !=null) {
122                             String newsResult= (String) msg.obj;
123                             List<NewsBean>nbl=JsonParse.getInstance().
124                                         getNewsList(newsResult);
125                             if(nbl !=null) {
126                                 if(nbl.size()>0) {
127                                     adapter.setData(nbl);
128                                 }
129                             }
130                         }
131                         break;
132                 }
133             }
134         }
135         class AdAutoSlidThread extends Thread {
```

```
136        @Override
137        public void run() {
138            super.run();
139            while (true) {
140                try {
141                    sleep(5000);
142                } catch(InterruptedException e) {
143                    e.printStackTrace();
144                }
145                if(mHandler !=null)
146                    mHandler.sendEmptyMessage(MSG_AD_SLID);
147            }
148        }
149    }
150    private void getNewsData() {
151        Request request=new Request.Builder().url(Constant.WEB_SITE+
152                    Constant.REQUEST_NEWS_URL).build();
153        Call call=okHttpClient.newCall(request);
154        //开启异步线程访问网络
155        call.enqueue(new Callback() {
156            @Override
157            public void onResponse(Response response) throws IOException {
158                String res=response.body().string();
159                Message msg=new Message();
160                msg.what=MSG_NEWS_OK;
161                msg.obj=res;
162                mHandler.sendMessage(msg);
163            }
164            @Override
165            public void onFailure(Request arg0, IOException arg1) {
166            }
167        });
168    }
169    private void getADData() {
170        Request request=new Request.Builder().url(Constant.WEB_SITE+
171                    Constant.REQUEST_AD_URL).build();
172        Call call=okHttpClient.newCall(request);
173        //开启异步线程访问网络
174        call.enqueue(new Callback() {
175            @Override
176            public void onResponse(Response response) throws IOException {
177                String res=response.body().string();
178                Message msg=new Message();
179                msg.what=MSG_AD_OK;
180                msg.obj=res;
181                mHandler.sendMessage(msg);
182            }
183            @Override
184            public void onFailure(Request arg0, IOException arg1) {
185            }
```

```
186             });
187         }
188     /**
189      * 计算控件大小
190      */
191     private void resetSize() {
192         int sw=UtilsHelper.getScreenWidth(getActivity());
193         int adLheight=sw / 2;          //广告条高度
194         ViewGroup.LayoutParams adlp=adBannerLay.getLayoutParams();
195         adlp.width=sw;
196         adlp.height=adLheight;
197         adBannerLay.setLayoutParams(adlp);
198     }
199 }
```

（5）修改 JsonParse.java 文件。由于首页中从服务器上获取的广告栏与新闻列表的数据不能直接加载到界面上，需要经过 gson 库解析后才能显示在界面上，因此需要在 com.itheima.topline.utils 包中的 JsonParse 类中添加如下代码：

```
1   public List<NewsBean>getAdList(String json) {
2       //使用 gson 库解析 JSON 数据
3       Gson gson=new Gson();
4       //创建一个 TypeToken 的匿名子类对象,并调用对象的 getType()方法
5       Type listType=new TypeToken<List<NewsBean>>() {
6       }.getType();
7       //把获取到的信息集合存到 adList 中
8       List<NewsBean>adList=gson.fromJson(json, listType);
9       return adList;
10  }
11  public List<NewsBean>getNewsList(String json) {
12      //使用 gson 库解析 JSON 数据
13      Gson gson=new Gson();
14      //创建一个 TypeToken 的匿名子类对象,并调用对象的 getType()方法
15      Type listType=new TypeToken<List<NewsBean>>() {
16      }.getType();
17      //把获取到的信息集合存到 newsList 中
18      List<NewsBean>newsList=gson.fromJson(json, listType);
19      return newsList;
20  }
```

（6）修改 AdBannerAdapter.java 文件。当用户触摸广告栏时，广告栏上的图片需要停止自动切换，因此需要找到文件 7-19 中的 onTouch()方法，在该方法中添加如下代码：

```
mHandler.removeMessages(HomeFragment.MSG_AD_SLID);
```

（7）修改底部导航栏。由于在点击底部导航栏的首页按钮时会出现首页界面，因此需要在第 6 章中找到文件 6-8，在该文件的 initView()方法中的"viewPager＝(ViewPager)findViewById(R.id.viewPager);"语句下方添加加载首页的代码，具体代码如下：

```
HomeFragment homeFragment=new HomeFragment();
List<Fragment>alFragment=new ArrayList<Fragment>();
alFragment.add(homeFragment);
//ViewPager 设置适配器
viewPager.setAdapter(new MyFragmentPagerAdapter(getSupportFragmentManager(),
        alFragment));
viewPager.setCurrentItem(0);          //ViewPager 显示第一个 Fragment
```

（8）创建 MyFragmentPagerAdapter.java 文件。由于首页界面用到了 ViewPager 控件，因此需要给该控件设置一个适配器，在 com.itheima.topline.adapter 文件夹中创建一个 MyFragmentPagerAdapter 类继承 FragmentPagerAdapter 类，具体代码如文件 7-22 所示。

【文件 7-22】 MyFragmentPagerAdapter.java

```
1   package com.itheima.topline.adapter;
2   import android.support.v4.app.Fragment;
3   import android.support.v4.app.FragmentManager;
4   import android.support.v4.app.FragmentPagerAdapter;
5   import java.util.List;
6   public class MyFragmentPagerAdapter extends FragmentPagerAdapter {
7       private List<Fragment>list;
8       public MyFragmentPagerAdapter(FragmentManager fm, List<Fragment>list) {
9           super(fm);
10          this.list=list;
11      }
12      @Override
13      public Fragment getItem(int position) {
14          return list.get(position);
15      }
16      @Override
17      public int getCount() {
18          return list.size();
19      }
20  }
```

7.4 新闻详情

任务综述

首页的广告栏图片与新闻列表的条目在被点击后会跳转到新闻详情界面，这个界面主要是通过 WebView 控件加载一个新闻链接展现界面信息的。该界面右上角有一个收藏图标，当用户处于登录状态时，点击收藏图标会把对应的新闻信息保存到数据库中，便于后续查询。当用户处于未登录状态时，点击收藏图标会提示用户"您还未登录，请先登录"（新闻收藏功能在实现登录注册后会进一步完善，此处暂不处理）。

【任务 7-15】 "新闻详情"界面

【任务分析】

在头条项目中,点击广告栏图片或新闻列表条目会跳转到"新闻详情"界面,"新闻详情"界面是通过 WebView 控件加载传递过来的新闻链接展示界面数据的,界面效果如图 7-5 所示。

图 7-5 "新闻详情"界面

【任务实施】

(1) 创建"新闻详情"界面。在 com.itheima.topline.activity 包中创建一个 Empty Activity 类,命名为 NewsDetailActivity,并将布局文件名指定为 activity_news_detail。

(2) 导入界面图片。将"新闻详情"界面所需的图片 collection_normal.png、collection_selected.png、app_loading_0.png、app_loading_1.png 导入 drawable-hdpi 文件夹。

(3) 放置界面控件。在布局文件中,放置一个 WebView 控件用于加载新闻地址;放置一个 TextView 控件用于显示 Now Loading 文字;一个 ProgressBar 控件用于显示正在加载的提示,具体代码如文件 7-23 所示。

【文件 7-23】 activity_news_detail.xml

```xml
1   <?xml version="1.0" encoding="utf-8"?>
2   <RelativeLayout xmlns:android="http://schemas.android.com/apk/res/android"
3       android:id="@+id/activity_news_detail"
4       android:layout_width="match_parent"
5       android:layout_height="match_parent">
6       <LinearLayout
7           android:layout_width="match_parent"
8           android:layout_height="match_parent"
9           android:orientation="vertical">
10          <include layout="@layout/main_titie_bar" />
11          <WebView
12              android:id="@+id/webView"
13              android:layout_width="fill_parent"
14              android:layout_height="fill_parent" />
15      </LinearLayout>
16      <LinearLayout
17          android:id="@+id/ll_loading"
18          android:layout_width="match_parent"
19          android:layout_height="match_parent"
20          android:gravity="center"
21          android:orientation="vertical"
22          android:visibility="gone">
23          <ProgressBar
24              android:layout_width="150dp"
25              android:layout_height="133dp"
26              android:indeterminateDrawable="@drawable/pb_loading" />
27          <TextView
28              android:layout_width="wrap_content"
29              android:layout_height="wrap_content"
30              android:text="Now Loading..."
31              android:textColor="@android:color/darker_gray" />
32      </LinearLayout>
33  </RelativeLayout>
```

（4）实现加载的动画效果。"新闻详情"界面在数据还未加载完全时会有一个提示正在加载数据的动画效果，这种效果可以通过逐帧动画实现。选中 drawable 文件夹，右击选择 New→Drawable resource file 选项，创建一个 pb_loading.xml 文件，在该文件中，android：drawable 表示需要显示的图片，android：duration 表示显示该图片的时间，具体代码如文件 7-24 所示。

【文件 7-24】 pb_loading.xml

```xml
1   <?xml version="1.0" encoding="UTF-8"?>
2   <animation-list android:oneshot="false"
3       xmlns:android="http://schemas.android.com/apk/res/android">
4       <item android:duration="150" android:drawable="@drawable/app_loading_0" />
5       <item android:duration="150" android:drawable="@drawable/app_loading_1" />
6   </animation-list>
```

(5) 修改 main_title_bar.xml 文件。根据"新闻详情"界面的效果可知,"新闻详情"界面标题栏的右边有一个收藏图标,因此需要在第 6 章中找到文件 6-3,在该文件中添加如下代码:

```xml
<ImageView
    android:id="@+id/iv_collection"
    android:layout_width="wrap_content"
    android:layout_height="wrap_content"
    android:layout_alignParentRight="true"
    android:layout_centerVertical="true"
    android:layout_marginRight="15dp"
    android:src="@drawable/collection_normal"
    android:visibility="gone" />
```

【任务 7-16】 "新闻详情"界面逻辑代码

【任务分析】

"新闻详情"界面主要加载一个从上个界面传递过来的新闻地址,展示"新闻详情"界面的信息。在"新闻详情"界面的右上角有一个收藏图标,当用户登录时,点击该图标会收藏本页面的新闻信息到数据库中,当用户未登录时,点击该图标会提示"您还未登录,请先登录"。

【任务实施】

(1) 获取界面控件。在 NewsDetailActivity 中创建界面控件的初始化方法 init(),用于获取新闻详情界面所要用到的控件。

(2) 设置 WebView 控件。在 NewsDetailActivity 中创建 WebView 控件的初始化方法 initWebView(),用于初始化 WebView 控件并设置新闻地址,具体代码如文件 7-25 所示。

【文件 7-25】 NewsDetailActivity.java

```
1   package com.itheima.topline.activity;
2   public class NewsDetailActivity extends AppCompatActivity {
3       private RelativeLayout rl_title_bar;
4       private WebView webView;
5       private TextView tv_main_title, tv_back;
6       private ImageView iv_collection;
7       private SwipeBackLayout layout;
8       private String newsUrl;
9       private NewsBean bean;
10      private String position;
11      private LinearLayout ll_loading;
12      @Override
13      protected void onCreate(Bundle savedInstanceState) {
14          super.onCreate(savedInstanceState);
15          layout=(SwipeBackLayout) LayoutInflater.from(this).inflate(
16              R.layout.base, null);
17          layout.attachToActivity(this);
18          setContentView(R.layout.activity_news_detail);
19          bean=(NewsBean) getIntent().getSerializableExtra("newsBean");
```

```java
20          position=getIntent().getStringExtra("position");
21          if(bean==null) return;
22          newsUrl=bean.getNewsUrl();
23          init();
24          initWebView();
25      }
26      private void init() {
27          tv_main_title=(TextView) findViewById(R.id.tv_main_title);
28          tv_main_title.setText("新闻详情");
29          rl_title_bar=(RelativeLayout) findViewById(R.id.title_bar);
30          rl_title_bar.setBackgroundColor(getResources().getColor(R.color.
31          rdTextColorPress));
32          ll_loading=(LinearLayout) findViewById(R.id.ll_loading);
33          iv_collection=(ImageView) findViewById(R.id.iv_collection);
34          iv_collection.setVisibility(View.VISIBLE);
35          tv_back=(TextView) findViewById(R.id.tv_back);
36          tv_back.setVisibility(View.VISIBLE);
37          tv_back.setOnClickListener(new View.OnClickListener() {
38              @Override
39              public void onClick(View view) {
40                  NewsDetailActivity.this.finish();
41              }
42          });
43          webView=(WebView) findViewById(R.id.webView);
44          iv_collection.setOnClickListener(new View.OnClickListener() {
45              @Override
46              public void onClick(View view) {
47              }
48          });
49      }
50      private void initWebView() {
51          webView.loadUrl(newsUrl);
52          WebSettings mWebSettings=webView.getSettings();
53          mWebSettings.setSupportZoom(true);
54          mWebSettings.setLoadWithOverviewMode(true);
55          mWebSettings.setUseWideViewPort(true);
56          mWebSettings.setDefaultTextEncodingName("GBK");      //设置解码格式
57          mWebSettings.setLoadsImagesAutomatically(true);
58          mWebSettings.setJavaScriptEnabled(true);             //支持 JavaScript 特效
59          //覆盖 WebView 默认使用第三方或系统默认浏览器打开网页的行为,使网页用 WebView 打开
60          webView.setWebViewClient(new WebViewClient() {
61              @Override
62              public boolean shouldOverrideUrlLoading(WebView view, String url) {
63                  //返回值是 true 时控制 WebView 打开,为 false 调用系统或第三方浏览器
64                  view.loadUrl(url);
65                  return true;
66              }
67              @Override
68              public void onPageStarted(WebView view, String url, Bitmap favicon) {
69                  super.onPageStarted(view, url, favicon);
```

```
70              ll_loading.setVisibility(View.VISIBLE);    //开始加载动画
71          }
72          @Override
73          public void onPageFinished(WebView view, String url) {
74              super.onPageFinished(view, url);
75              ll_loading.setVisibility(View.GONE);        //当加载结束时隐藏动画
76          }
77      });
78  }
79 }
```

（3）自定义 SwipeBackLayout 控件。在头条项目中，向右滑动"新闻详情"界面会关闭该界面，因此需要通过自定义控件 SwipeBackLayout 完成。首先将图片 shadow_left.png 导入 drawable-hdpi 文件夹，然后在 com.itheima.topline.view 包中创建一个 SwipeBackLayout 类并继承 FrameLayout 类，具体代码如文件 7-26 所示。

【文件 7-26】 SwipeBackLayout.java

```
1  package com.itheima.topline.view;
2  public class SwipeBackLayout extends FrameLayout {
3      private View mContentView;
4      private int mTouchSlop,downX,downY,tempX,viewWidth;
5      private Scroller mScroller;
6      private boolean isSilding,isFinish;
7      private Drawable mShadowDrawable;
8      private Activity mActivity;
9      private List<ViewPager>mViewPagers=new LinkedList<ViewPager>();
10     public SwipeBackLayout(Context context, AttributeSet attrs) {
11         this(context, attrs, 0);
12     }
13     public SwipeBackLayout(Context context, AttributeSet attrs, int defStyle) {
14         super(context, attrs, defStyle);
15         //触发移动事件的最短距离,如果小于这个距离则不触发移动控件
16         mTouchSlop=ViewConfiguration.get(context).getScaledTouchSlop();
17         mScroller=new Scroller(context);
18         mShadowDrawable=getResources().getDrawable(R.drawable.shadow_left);
19     }
20     public void attachToActivity(Activity activity) {
21         mActivity=activity;
22         TypedArray a=activity.getTheme().obtainStyledAttributes(
23             new int[]{android.R.attr.windowBackground});
24         int background=a.getResourceId(0, 0);
25         a.recycle();
26         ViewGroup decor= (ViewGroup) activity.getWindow().getDecorView();
27         ViewGroup decorChild= (ViewGroup) decor.getChildAt(0);
28         decorChild.setBackgroundResource(background);
29         decor.removeView(decorChild);
30         addView(decorChild);
31         setContentView(decorChild);
```

```
32              decor.addView(this);
33          }
34          private void setContentView(View decorChild) {
35              mContentView= (View) decorChild.getParent();
36          }
37          /**
38           * 事件拦截操作
39           */
40          @Override
41          public boolean onInterceptTouchEvent(MotionEvent ev) {
42              //处理 ViewPager 冲突问题
43              ViewPager mViewPager=getTouchViewPager(mViewPagers, ev);
44              if(mViewPager !=null && mViewPager.getCurrentItem() !=0) {
45                  return super.onInterceptTouchEvent(ev);
46              }
47              switch(ev.getAction()) {
48                  case MotionEvent.ACTION_DOWN:
49                      downX=tempX= (int) ev.getRawX();
50                      downY= (int) ev.getRawY();
51                      break;
52                  case MotionEvent.ACTION_MOVE:
53                      int moveX= (int) ev.getRawX();
54                      //满足此条件屏蔽 SildingFinishLayout 中子类的 touch 事件
55                      if(moveX -downX>mTouchSlop
56                              && Math.abs((int) ev.getRawY() -downY)<mTouchSlop) {
57                          return true;
58                      }
59                      break;
60              }
61              return super.onInterceptTouchEvent(ev);
62          }
63          @Override
64          public boolean onTouchEvent(MotionEvent event) {
65              switch(event.getAction()) {
66                  case MotionEvent.ACTION_MOVE:
67                      int moveX= (int) event.getRawX();
68                      int deltaX=tempX -moveX;
69                      tempX=moveX;
70                      if(moveX -downX>mTouchSlop
71                              && Math.abs((int) event.getRawY() -downY)<mTouchSlop) {
72                          isSilding=true;
73                      }
74                      if(moveX -downX>=0 && isSilding) {
75                          mContentView.scrollBy(deltaX, 0);
76                      }
77                      break;
78                  case MotionEvent.ACTION_UP:
79                      isSilding=false;
80                      if(mContentView.getScrollX()<=-viewWidth / 2) {
81                          isFinish=true;
```

```
82                    scrollRight();
83                } else {
84                    scrollOrigin();
85                    isFinish=false;
86                }
87                break;
88            }
89        return true;
90    }
91    /**
92     * 获取 SwipeBackLayout 中 ViewPager 的集合
93     */
94    private void getAlLViewPager(List<ViewPager>mViewPagers, ViewGroup parent) {
95        int childCount=parent.getChildCount();
96        for(int i=0; i<childCount; i++) {
97            View child=parent.getChildAt(i);
98            if(child instanceof ViewPager) {
99                mViewPagers.add((ViewPager) child);
100           } else if(child instanceof ViewGroup) {
101               getAlLViewPager(mViewPagers, (ViewGroup) child);
102           }
103       }
104   }
105   /**
106    * 返回 touch 的 ViewPager
107    */
108   private ViewPager getTouchViewPager(List<ViewPager>mViewPagers,
109   MotionEvent ev){
110       if(mViewPagers==null || mViewPagers.size()==0) {
111           return null;
112       }
113       Rect mRect=new Rect();
114       for(ViewPager v : mViewPagers) {
115           v.getHitRect(mRect);
116           if(mRect.contains((int) ev.getX(), (int) ev.getY())) {
117               return v;
118           }
119       }
120       return null;
121   }
122   @Override
123   protected void onLayout(boolean changed, int l, int t, int r, int b) {
124       super.onLayout(changed, l, t, r, b);
125       if(changed) {
126           viewWidth=this.getWidth();
127           getAlLViewPager(mViewPagers, this);
128       }
129   }
130   @Override
131   protected void dispatchDraw(Canvas canvas) {
```

```
132         super.dispatchDraw(canvas);
133         if(mShadowDrawable !=null && mContentView !=null) {
134             int left=mContentView.getLeft()-mShadowDrawable.getIntrinsicWidth();
135             int right=left+mShadowDrawable.getIntrinsicWidth();
136             int top=mContentView.getTop();
137             int bottom=mContentView.getBottom();
138             mShadowDrawable.setBounds(left, top, right, bottom);
139             mShadowDrawable.draw(canvas);
140         }
141     }
142     /**
143      * 滚动出界面
144      */
145     private void scrollRight() {
146         final int delta=(viewWidth+mContentView.getScrollX());
147         //调用 startScroll 方法设置一些滚动的参数
148         mScroller.startScroll(mContentView.getScrollX(), 0, -delta+1, 0,
149             Math.abs(delta));
150         postInvalidate();
151     }
152     /**
153      * 滚动到起始位置
154      */
155     private void scrollOrigin() {
156         int delta=mContentView.getScrollX();
157         mScroller.startScroll(mContentView.getScrollX(), 0, -delta, 0,
158             Math.abs(delta));
159         postInvalidate();
160     }
161     @Override
162     public void computeScroll() {
163         //调用 startScroll()方法时,scroller.computeScrollOffset()返回为 true,
164         if(mScroller.computeScrollOffset()) {
165             mContentView.scrollTo(mScroller.getCurrX(), mScroller.getCurrY());
166             postInvalidate();
167             if(mScroller.isFinished() && isFinish) {
168                 mActivity.finish();
169             }
170         }
171     }
172 }
```

（4）创建 base.xml 文件。由于需要把自定义控件 SwipeBackLayout 添加到"新闻详情"界面中,因此需要在 res/layout 文件夹中创建一个 base.xml 文件,在该文件中引入 SwipeBackLayout 自定义控件即可,具体代码如文件 7-27 所示。

【文件 7-27】 base.xml

```
1 <com.itheima.topline.view.SwipeBackLayout
2     xmlns:android="http://schemas.android.com/apk/res/android"
```

```
3       android:layout_width="match_parent"
4       android:layout_height="match_parent">
5  </com.itheima.topline.view.SwipeBackLayout>
```

注意：如果需要向右滑动关闭一个界面，则需要添加一个自定义控件 SwipeBackLayout，并把添加该控件的语句放在该界面所对应的 Activity 中的 setContentView()方法之前，添加该自定义控件的代码如下所示：

```
layout= (SwipeBackLayout) LayoutInflater.from(this).inflate(R.layout.base, null);
layout.attachToActivity(this);
```

（5）修改 AdBannerFragment 类。由于点击首页的广告栏图片会跳转到"新闻详情"界面，因此需要找到文件 7-18 中的 onCreateView()方法，在该方法的图片点击事件中添加如下代码：

```
if(nb==null) return;
Intent intent=new Intent(getActivity(), NewsDetailActivity.class);
intent.putExtra("newsBean", nb);
getActivity().startActivity(intent);
```

（6）修改 HomeListAdapter 类。由于点击首页的新闻列表的条目会跳转到"新闻详情"界面，因此需要找到文件 7-20 中的 onBindViewHolder()方法，在该方法中的 itemView 点击事件中添加如下代码：

```
Intent intent=new Intent(context, NewsDetailActivity.class);
intent.putExtra("newsBean", bean);
context.startActivity(intent);
```

（7）修改清单文件。由于在向右滑动"新闻详情"界面时，在关闭该界面之前会出现空白界面，因此需要在清单文件中把"新闻详情"界面的主题修改为透明主题。首先在 res/values 文件夹的 styles.xml 文件中添加一个名为 AppTheme.TransparentActivity 的透明主题，具体代码如下：

```
<style name="AppTheme.TransparentActivity" parent="Theme.AppCompat.NoActionBar">
  <item name="android:windowBackground">@android:color/transparent</item>
  <item name="android:windowIsTranslucent">true</item>
</style>
```

然后在清单文件的 NewsDetailActivity 中引用此透明主题，具体代码如下：

```
<activity
    android:name=".activity.NewsDetailActivity"
    android:theme="@style/AppTheme.TransparentActivity" />
```

由于"新闻详情"界面需要加载网页，因此需要在清单文件中添加网络权限，具体代码如下：

```
<uses-permission android:name="android.permission.INTERNET"/>
```

7.5 Python 学科

任务综述

在头条项目中,点击首页中的四个学科(Python、Java、PHP、Android)按钮,分别会跳转到对应学科的界面,该界面展示的是学科在全国各个分校的开班地址和开班情况等信息,各个学科界面的数据以 JSON 格式存放在 Tomcat 服务器,在各个界面的逻辑代码中,调用 gson 库解析获取的 JSON 数据并展示在对应学科的界面上。

【任务 7-17】 "Python 学科"界面

【任务分析】

在头条项目中,点击首页中的四个学科(Python、Java、PHP、Android)按钮,分别会跳转到对应学科的界面,该界面展示的是对应学科在全国各个分校的开班地址和开班情况等信息,此处以 Python 学科为例,界面效果如图 7-6 所示。

图 7-6 "Python 学科"界面

【任务实施】

(1)创建"Python 学科"界面。在 com.itheima.topline.activity 包中创建一个 Empty Activity 类,命名为 PythonActivity,该界面与首页界面新闻列表的布局类似,因此该界面调用首页界面的布局文件 fragment_home.xml。

(2)导入界面图片。将"Python 学科"界面所需的图片 fire_icon.png 导入 drawable-hdpi 文件夹。

(3)修改 fragment_home.xml 文件。由于 Python 学科界面设有标题栏,而 fragment_home.xml 文件中只有下拉刷新与 WrapRecyclerView 列表,因此需要在该文件的 com.itheima.PullToRefreshView 控件上方引入标题栏,具体代码如下:

```
<include layout="@layout/main_title_bar" />
```

(4)修改 HomeFragment 类。由于 fragment_home.xml 文件引入了标题栏,但是首页界面的新闻列表不能显示标题栏,因此需要在首页界面的逻辑代码中把标题栏设置为不可见的状态,找到文件 7-21,在该文件的"private HomeListAdapter adapter;"语句下方添加如下代码:

```
private RelativeLayout rl_title_bar;
```

在该文件的 initView()方法中的"View view = inflater.inflate(R.layout.fragment_ home，container，false）；"语句下方添加如下代码：

```
rl_title_bar=(RelativeLayout) view.findViewById(R.id.title_bar);
rl_title_bar.setVisibility(View.GONE);
```

【任务 7-18】 "Python 学科"界面 Item

【任务分析】

"Python 学科"界面通过使用 WrapRecyclerView 控件展示开班信息列表，因此需要创建一个该列表的 Item 界面。在 Item 界面中需要展示开班地址、开班内容以及"火"的图标，界面效果如图 7-7 所示。

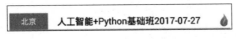

图 7-7 "Python 学科"界面 Item

【任务实施】

（1）创建"Python 学科"界面 Item。在 res/layout 文件夹中创建布局文件 python_list_item.xml。

（2）放置界面控件。在布局文件中，放置两个 TextView 控件分别用于显示分校地址与开班情况；一个 ImageView 控件用于显示"火"的图标，一个 View 控件用于显示一条灰色的线，具体代码如文件 7-28 所示。

【文件 7-28】 python_list_item.xml

```
1   <?xml version="1.0" encoding="utf-8"?>
2   <RelativeLayout xmlns:android="http://schemas.android.com/apk/res/android"
3       android:layout_width="match_parent"
4       android:layout_height="40dp"
5       android:background="@android:color/white"
6       android:paddingLeft="8dp"
7       android:paddingRight="8dp">
8       <TextView
9           android:id="@+id/tv_address"
10          android:layout_width="60dp"
11          android:layout_height="25dp"
12          android:layout_centerVertical="true"
13          android:background="@color/rdTextColorPress"
14          android:gravity="center"
15          android:textColor="@android:color/white"
16          android:textSize="12sp" />
17      <TextView
18          android:id="@+id/tv_content"
19          android:layout_width="wrap_content"
20          android:layout_height="wrap_content"
21          android:layout_centerVertical="true"
```

```xml
22        android:layout_marginLeft="15dp"
23        android:layout_toRightOf="@id/tv_address"
24        android:textColor="@android:color/black"
25        android:textSize="14sp" />
26    <ImageView
27        android:id="@+id/iv_fire"
28        android:layout_width="20dp"
29        android:layout_height="20dp"
30        android:layout_alignParentRight="true"
31        android:layout_centerVertical="true"
32        android:scaleType="fitXY"
33        android:src="@drawable/fire_icon" />
34    <View
35        android:layout_width="match_parent"
36        android:layout_height="1dp"
37        android:layout_below="@id/tv_address"
38        android:background="@android:color/darker_gray" />
39 </RelativeLayout>
```

【任务 7-19】 创建 PythonBean

【任务分析】

由于"Python 学科"界面的每个 Item 中都有开班 Id、开班地址、开班内容等属性，因此可以创建一个 PythonBean 类存放这些属性。

【任务实施】

在 com.itheima.topline.bean 包中创建一个 PythonBean 类，在该类中创建 Python 学科开班信息的所有属性，具体代码如文件 7-29 所示。

【文件 7-29】 PythonBean.java

```java
1  package com.itheima.topline.bean;
2  public class PythonBean {
3      private int id;                  //开班 Id
4      private String address;          //开班地址
5      private String content;          //开班内容
6      public int getId() {
7          return id;
8      }
9      public void setId(int id) {
10         this.id=id;
11     }
12     public String getAddress() {
13         return address;
14     }
15     public void setAddress(String address) {
16         this.address=address;
17     }
```

```
18      public String getContent() {
19          return content;
20      }
21      public void setContent(String content) {
22          this.content=content;
23      }
24  }
```

【任务 7-20】 "Python 学科"界面 Adapter

【任务分析】

"Python 学科"界面的开班信息列表是通过 WrapRecyclerView 控件展示的,因此需要创建一个数据适配器 PythonListAdapter 对 WrapRecyclerView 控件进行数据适配。

【任务实施】

(1) 创建 PythonListAdapter 类。在 com.itheima.topline.adapter 包中创建一个 PythonListAdapter 类继承 RecyclerView.Adapter＜RecyclerView.ViewHolder＞类,并重写 onCreateViewHolder()、onBindViewHolder()、getItemCount()方法。

(2) 创建 PythonViewHolder 类。在 PythonListAdapter 类中创建一个 PythonViewHolder 类以获取 Item 界面上的控件,具体代码如文件 7-30 所示。

【文件 7-30】 PythonListAdapter.java

```
1   package com.itheima.topline.adapter;
2   public class PythonListAdapter extends RecyclerView.Adapter<RecyclerView.
3   ViewHolder>{
4       private List<PythonBean>pbl;
5       public void setData(List<PythonBean>pbl) {
6           this.pbl=pbl;
7           notifyDataSetChanged();
8       }
9       @Override
10      public RecyclerView.ViewHolder onCreateViewHolder(ViewGroup viewGroup,
11      int viewType) {
12          View view=LayoutInflater.from(viewGroup.getContext()).inflate(
13                              R.layout.python_list_item, viewGroup, false);
14          PythonViewHolder viewHolder=new PythonViewHolder(view);
15          return viewHolder;
16      }
17      @Override
18      public void onBindViewHolder(final RecyclerView.ViewHolder holder, int i)
19      {
20          final PythonBean bean=pbl.get(i);
21          ((PythonViewHolder) holder).tv_address.setText(bean.getAddress());
22          ((PythonViewHolder) holder).tv_content.setText(bean.getContent());
23      }
24      @Override
```

```
25    public int getItemCount() {
26        return pbl==null ? 0 : pbl.size();
27    }
28    public class PythonViewHolder extends RecyclerView.ViewHolder {
29        private TextView tv_address, tv_content;
30        public PythonViewHolder(View itemView) {
31            super(itemView);
32            tv_address= (TextView) itemView.findViewById(R.id.tv_address);
33            tv_content= (TextView) itemView.findViewById(R.id.tv_content);
34        }
35    }
36 }
```

【任务 7-21】 "Python 学科"界面逻辑代码

【任务分析】

"Python 学科"界面上加载的数据需要在 Tomcat 服务器上设置,同时需要在 JsonParse 类中创建 getPythonList()方法用于解析从服务器上获取的 JSON 格式的数据。

【任务实施】

(1) 创建"Python 学科"界面数据文件。在 Tomcat 的 ROOT/topline 目录中创建一个 python_list_data.json 文件,该文件用于存放 Python 界面需要加载的数据,具体代码如文件 7-31 所示。

【文件 7-31】 python_list_data.json

```
1  [
2    {
3      "id":1,
4      "address":"北京",
5      "content":"人工智能+Python 基础班 2017-08-18",
6      "open_class":"我要报名"
7    },
8    {
9      "id":2,
10     "address":"北京",
11     "content":"人工智能+Python 基础班 2017-07-27",
12     "open_class":"我要报名"
13   },
14   ...由于数据量太大,此处省略一部分数据
15   {
16     "id":23,
17     "address":"在线学习",
18     "content":"人工智能+Python 在线班 2017-06-29",
19     "open_class":"我要报名"
20   }
21 ]
```

（2）在 Constant.java 文件中添加请求地址。由于"Python 学科"界面的数据是从服务器获取的，因此需要找到文件 7-3，在该文件中添加"Python 学科"界面请求数据需要的地址，具体代码如下：

```java
public static final String REQUEST_PYTHON_URL="/python_list_data.json";
```

（3）解析 JSON 数据。由于从 Tomcat 服务器上获取的 JSON 格式的数据不能直接加载到界面上，因此需要在 com.itheima.topline.utils 包的 JsonParse 类中创建一个 getPythonList() 方法，用于解析该界面获取的 JSON 数据，在 JsonParse 类中需添加如下代码：

```java
1  public List<PythonBean>getPythonList(String json) {
2      //使用 gson 库解析 JSON 数据
3      Gson gson=new Gson();
4      //创建一个 TypeToken 的匿名子类对象，并调用对象的 getType()方法
5      Type listType=new TypeToken<List<PythonBean>>() {}.getType();
6      //把获取到的信息集合存到 pythonList 中
7      List<PythonBean>pythonList=gson.fromJson(json, listType);
8      return pythonList;
9  }
```

（4）获取数据。在 PythonActivity 中创建一个 initData() 方法，用于获取"Python 学科"界面需要加载的数据。

（5）获取界面控件。在 PythonActivity 中创建界面控件的初始化方法 initView()，用于获取"Python 学科"界面所要用到的控件，具体代码如文件 7-32 所示。

【文件 7-32】 PythonActivity.java

```java
1  package com.itheima.topline.activity;
2  public class PythonActivity extends AppCompatActivity {
3      private SwipeBackLayout layout;
4      private PullToRefreshView mPullToRefreshView;
5      private WrapRecyclerView recyclerView;
6      public static final int REFRESH_DELAY=1000;
7      public static final int MSG_PYTHON_OK=1;          //获取数据
8      private TextView tv_main_title, tv_back;
9      private RelativeLayout rl_title_bar;
10     private MHandler mHandler;
11     private PythonListAdapter adapter;
12     @Override
13     protected void onCreate(Bundle savedInstanceState) {
14         super.onCreate(savedInstanceState);
15         layout=(SwipeBackLayout) LayoutInflater.from(this).inflate(
16             R.layout.base, null);
17         layout.attachToActivity(this);
18         setContentView(R.layout.fragment_home);
19         mHandler=new MHandler();
```

```java
20        initData();
21        initView();
22    }
23    private void initView() {
24        tv_main_title= (TextView) findViewById(R.id.tv_main_title);
25        tv_main_title.setText("Python学科");
26        rl_title_bar= (RelativeLayout) findViewById(R.id.title_bar);
27        rl_title_bar.setBackgroundColor(getResources().getColor(
28                                        R.color.rdTextColorPress));
29        tv_back= (TextView) findViewById(R.id.tv_back);
30        tv_back.setVisibility(View.VISIBLE);
31        recyclerView= (WrapRecyclerView) findViewById(R.id.recycler_view);
32        recyclerView.setLayoutManager(new LinearLayoutManager(this));
33        adapter=new PythonListAdapter();
34        recyclerView.setAdapter(adapter);
35        mPullToRefreshView= (PullToRefreshView) findViewById(R.id.pull_to_refresh);
36        mPullToRefreshView.setOnRefreshListener(new PullToRefreshView.
37                                        OnRefreshListener() {
38            @Override
39            public void onRefresh() {
40                mPullToRefreshView.postDelayed(new Runnable() {
41                    @Override
42                    public void run() {
43                        mPullToRefreshView.setRefreshing(false);
44                        initData();
45                    }
46                }, REFRESH_DELAY);
47            }
48        });
49        tv_back.setOnClickListener(new View.OnClickListener() {
50            @Override
51            public void onClick(View view) {
52                PythonActivity.this.finish();
53            }
54        });
55    }
56    private void initData() {
57        OkHttpClient okHttpClient=new OkHttpClient();
58        Request request=new Request.Builder().url(Constant.WEB_SITE+
59                                        Constant.REQUEST_PYTHON_URL).build();
60        Call call=okHttpClient.newCall(request);
61        //开启异步线程访问网络
62        call.enqueue(new Callback() {
63            @Override
64            public void onResponse(Response response) throws IOException {
65                String res=response.body().string();
66                Message msg=new Message();
67                msg.what=MSG_PYTHON_OK;
```

```
68                msg.obj=res;
69                mHandler.sendMessage(msg);
70            }
71            @Override
72            public void onFailure(Request arg0, IOException arg1) {
73            }
74        });
75    }
76    /**
77     * 事件捕获
78     */
79    class MHandler extends Handler {
80        @Override
81        public void dispatchMessage(Message msg) {
82            super.dispatchMessage(msg);
83            switch(msg.what) {
84                case MSG_PYTHON_OK:
85                    if(msg.obj !=null) {
86                        String vlResult= (String) msg.obj;
87                        //使用 Gson 解析数据
88                        List<PythonBean>pythonList=JsonParse.getInstance().
89                                                  getPythonList(vlResult);
90                        adapter.setData(pythonList);
91                    }
92                    break;
93            }
94        }
95    }
96 }
```

（6）修改 HomeFragment.java 文件。由于点击首页界面的"Python 学科"按钮会跳转到"Python 学科"界面，因此需要找到文件 7-21 中的 setListener( )方法，在该方法的 onClick( )方法中添加如下代码：

```
Intent intent=new Intent(getActivity(), PythonActivity.class);
startActivity(intent);
```

（7）修改清单文件。由于"Python 学科"界面向右滑动会关闭该界面，因此需要给该界面添加透明主题样式，在清单文件中的 PythonActivity 对应的 activity 标签中添加如下代码：

```
android:theme="@style/AppTheme.TransparentActivity"
```

7.6 本章小结

本章主要讲解了头条项目的首页模块，首页模块主要包括搭建服务器、工具类、首页、新闻详情以及 Python 学科。首页界面、"Python 学科"界面的数据获取与数据解析逻辑流程较为复杂,读者需要认真阅读和仔细分析。

【思考题】

1. 如何解析 JSON 文件？
2. 如何实现首页界面与 Python 界面的数据展示？

第 8 章 统 计 模 块

学习目标
- 掌握"统计"界面的开发,学会使用第三方圆形菜单
- 掌握"统计详情"界面的开发,使用 HelloCharts 实现统计功能

统计模块主要用于展示一个炫酷的圆形菜单,每个菜单代表一个学科,当点击对应学科时会通过饼状图、柱状图、线形图等图表形式直观地展示该学科的统计信息,该统计模块通过开源图表库 HelloCharts 实现,本章将针对统计模块进行详细讲解。

8.1 统计

任务综述

思政材料 8

"统计"界面主要展示一个圆形菜单,当点击圆形菜单时会弹出 9 个学科的菜单并展示在"统计"界面上,点击任意一个菜单都可以跳转到对应学科的"统计详情"界面。

【任务 8-1】 "统计"界面

【任务分析】

"统计"界面主要展示一个圆形菜单,点击圆形菜单会出现 9 个学科的菜单,界面效果如图 8-1 所示。

【任务实施】

(1) 创建"统计"界面。在 res/layout 文件夹中创建一个布局文件 fragment_count.xml。

(2) 导入界面图片。将"统计"界面所需的图片 bat.png、bear.png、bee.png、butterfly.png、cat.png、deer.png、dolphin.png、eagle.png、horse.png、elephant.png、owl.png、peacock.png、pig.png、rat.png、snake.png、squirrel.png、count_bg.png 导入 drawable-hdpi 文件夹。

(3) 引入第三方酷炫菜单。在"统计"界面中,右下角的圆形菜单通过引入第三方框架实现,在 Android Studio 中,选择 File→New→Import Module 选项,把第三方库 boommenu 导入项目中,选中项目,右击选择 Open Module Settings 选项后选择 Dependencies 选项卡,单击右上角的绿色加号并选择 Module Dependency 选项,把 boommenu 框架加入主项目,圆形菜单框架详情如图 8-2 所示。

图 8-1 "统计"界面

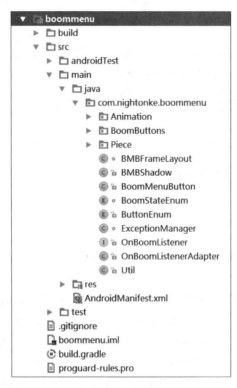

图 8-2 圆形菜单框架

（4）放置界面控件。在布局文件中，放置一个 BoomMenuButton 控件用于显示圆形菜单，具体代码如文件 8-1 所示。

【文件 8-1】 fragment_count.xml

```xml
1  <?xml version="1.0" encoding="utf-8"?>
2  <RelativeLayout xmlns:android="http://schemas.android.com/apk/res/android"
3      xmlns:app="http://schemas.android.com/apk/res-auto"
4      android:layout_width="match_parent"
5      android:layout_height="match_parent"
6      android:background="@drawable/count_bg">
7      <com.nightonke.boommenu.BoomMenuButton
8          android:id="@+id/bmb"
9          android:layout_width="wrap_content"
10         android:layout_height="wrap_content"
11         android:layout_alignParentBottom="true"
12         android:layout_alignParentRight="true"
13         app:bmb_highlightedColor="@android:color/holo_blue_dark"
14         app:bmb_normalColor="@android:color/holo_blue_light" />
15 </RelativeLayout>
```

【任务 8-2】 "统计"界面逻辑代码

【任务分析】

"统计"界面的右下角展示了一个圆形菜单,点击该菜单会动态地弹出 9 个学科菜单,点击任意一个学科都可以进入对应学科的"统计详情"界面。

【任务实施】

（1）获取界面控件。在 com.itheima.topline.fragment 包中创建一个 CountFragment 类继承 Fragment 类,并在该类中重写 onCreateView()方法,用于加载布局文件以及设置菜单的样式。

（2）设置菜单按钮的点击事件。在 CountFragment 中创建一个 addBuilder()方法,用于处理 9 个学科按钮的点击事件,此处以前两个按钮（Android 学科、Java 学科）为例,具体代码如文件 8-2 所示。

【文件 8-2】 CountFragment.java

```java
1  package com.itheima.topline.fragment;
2  public class CountFragment extends Fragment {
3      private BoomMenuButton bmb;
4      public CountFragment() {
5      }
6      @Override
7      public View onCreateView(LayoutInflater inflater, ViewGroup container,
8                               Bundle savedInstanceState) {
9          View view=inflater.inflate(R.layout.fragment_count, container, false);
10         bmb= (BoomMenuButton) view.findViewById(R.id.bmb);
```

```
11          assert bmb !=null;
12          //设置点击圆形菜单后显示的多个按钮为圆形且带文本
13          bmb.setButtonEnum(ButtonEnum.TextInsideCircle);
14          bmb.setPiecePlaceEnum(PiecePlaceEnum.DOT_9_1);        //设置右下角圆形菜单中有9个圆形
15          //设置点击右下角圆形菜单后显示的按钮为9个圆形 Button
16          bmb.setButtonPlaceEnum(ButtonPlaceEnum.SC_9_1);
17          for(int i=0; i<bmb.getPiecePlaceEnum().pieceNumber(); i++) {
18              addBuilder();
19          }
20          return view;
21      }
22      private void addBuilder() {
23          bmb.addBuilder(new TextInsideCircleButton.Builder()
24                  .normalImageRes(BuilderManager.getImageResource())
25                  .normalTextRes(BuilderManager.getTextResource())
26                  .listener(new OnBMClickListener() {
27                      @Override
28                      public void onBoomButtonClick(int index) {
29                          switch(index) {
30                              case 0:        //跳转到 Android 统计详情界面
31                                  break;
32                              case 1:        //跳转到 Java 统计详情界面
33                                  break;
34                          }
35                      }
36                  }));
37      }
38  }
```

（3）创建一个 BuilderManager 类。由于点击圆形菜单后会显示9个学科的按钮，这9个按钮的文本与背景图片需要设置成统计界面展现的样式，因此需要在 com.itheima.topline.utils 包中创建一个 BuilderManager 类以实现这些功能，具体代码如文件 8-3 所示。

【文件 8-3】 BuilderManager.java

```
1   package com.itheima.topline.utils;
2   public class BuilderManager {
3       private static int[] imageResources=new int[]{        //9个菜单随机选择的图片
4               R.drawable.bat,
5               R.drawable.bear,
6               R.drawable.bee,
7               R.drawable.butterfly,
8               R.drawable.cat,
9               R.drawable.deer,
10              R.drawable.dolphin,
11              R.drawable.eagle,
12              R.drawable.horse,
13              R.drawable.elephant,
14              R.drawable.owl,
15              R.drawable.peacock,
```

```
16                R.drawable.pig,
17                R.drawable.rat,
18                R.drawable.snake,
19                R.drawable.squirrel
20        };
21        private static int[] textResources=new int[]{        //9个菜单中的文本
22                R.string.android,
23                R.string.java,
24                R.string.python,
25                R.string.php,
26                R.string.c,
27                R.string.ios,
28                R.string.fore_end,
29                R.string.ui,
30                R.string.network
31        };
32        private static int imageResourceIndex=0;
33        private static int textResourceIndex=0;
34        public static int getImageResource() {
35            if(imageResourceIndex>=imageResources.length) imageResourceIndex=0;
36            return imageResources[imageResourceIndex++];
37        }
38        public static int getTextResource() {
39            if(textResourceIndex>=textResources.length) textResourceIndex=0;
40            return textResources[textResourceIndex++];
41        }
42        private static BuilderManager ourInstance=new BuilderManager();
43        public static BuilderManager getInstance() {
44            return ourInstance;
45        }
46        private BuilderManager() {
47        }
48  }
```

（4）修改 strings.xml 文件。由于"统计"界面中的 9 个学科的文本信息量比较多，因此为了方便，可以把这 9 个文本写在 res/values 文件夹的 strings.xml 文件中，具体代码如下：

```
<string name="android">黑马程序员.Android</string>
<string name="java">黑马程序员.Java</string>
<string name="php">黑马程序员.PHP</string>
<string name="python">黑马程序员.Python</string>
<string name="c">黑马程序员.C/C++</string>
<string name="ios">黑马程序员.iOS</string>
<string name="fore_end">黑马程序员.前端与移动开发</string>
<string name="ui">黑马程序员.UI 设计</string>
<string name="network">黑马程序员.网络营销</string>
```

（5）修改底部导航栏。由于点击底部导航栏的"统计"按钮会出现"统计"界面，因此需要找到第 6 章中的文件 6-8，在该文件的 initView() 方法中的"HomeFragment homeFragment＝new HomeFragment();"语句下方添加如下代码：

```
CountFragment countFragment=new CountFragment();
```

在"alFragment.add(homeFragment);"语句下方添加如下代码:

```
alFragment.add(countFragment);
```

8.2 统计详情

任务综述

"统计详情"界面主要使用第三方图表库 HelloCharts,通过饼状图、折线图以及柱状图展示该学科的市场薪资比例,本模块以 Java 学科和 Android 学科为例进行演示。

【任务 8-3】 "Android 统计"详情界面

【任务分析】

"Android 统计"详情界面主要以饼状图的形式展示 Android 薪资的占比情况,界面效果如图 8-3 所示。

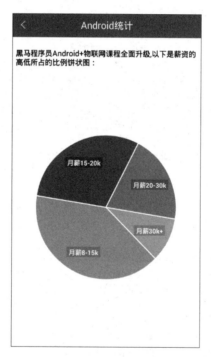

图 8-3 "Android 统计"详情界面

【任务实施】

(1) 创建"Android 统计"详情界面。在 com.itheima.topline.activity 包中创建一个 Empty Activity 类,命名为 AndroidCountActivity 并将布局文件名指定为 activity_android_count。

(2) 引入第三方库 hellocharts-library。在"Android 统计"详情界面中,饼状图是通过引入第三方库完成的。在 Android Studio 中,选择 File→New→Import Module 选项,把第三方库 hellocharts-library 导入项目,选中项目,右击选择 Open Module Settings 选项后选择 Dependencies 选项卡,单击右上角的绿色加号并选择 Module Dependency 选项,把 hellocharts-library 库加入主项目,hellocharts-library 库的详情如图 8-4 所示。

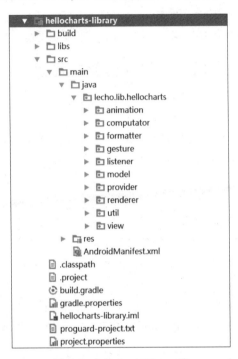

图 8-4　hellocharts-library 库

(3) 放置界面控件。在布局文件中,放置一个 TextView 控件用于显示信息的文本;一个 PieChartView 控件用于显示饼状图,具体代码如文件 8-4 所示。

【文件 8-4】　activity_android_count.xml

```
1   <?xml version="1.0" encoding="utf-8"?>
2   <LinearLayout xmlns:android="http://schemas.android.com/apk/res/android"
3       android:layout_width="match_parent"
4       android:layout_height="match_parent"
5       android:background="@android:color/white"
6       android:orientation="vertical">
7       <include layout="@layout/main_title_bar" />
8       <TextView
9           android:id="@+id/tv_intro"
10          android:layout_width="match_parent"
11          android:layout_height="wrap_content"
12          android:layout_marginLeft="8dp"
13          android:layout_marginRight="8dp"
14          android:layout_marginTop="20dp"
15          android:textColor="@android:color/black"
```

```
16              android:textSize="14sp" />
17      <lecho.lib.hellocharts.view.PieChartView
18          android:id="@+id/chart"
19          android:layout_width="match_parent"
20          android:layout_height="match_parent"
21          android:layout_marginLeft="40dp"
22          android:layout_marginRight="40dp"/>
23  </LinearLayout>
```

【任务 8-4】 "Android 统计"详情界面逻辑代码

【任务分析】

"Android 统计"详情界面主要展示一个薪资占比的饼状图,该饼状图是通过 HelloCharts 库中的 PieChartView 控件实现的。

【任务实施】

(1) 获取界面控件。在 AndroidCountActivity 中创建界面控件的初始化方法 init(),用于获取"Android 统计"详情界面所要用到的控件。

(2) 设置饼状图中的文字。在 AndroidCountActivity 中创建一个 toggleLabels()方法,用于设置饼状图中的文字效果。

(3) 设置饼状图中的扇形。由于饼状图中有 4 个颜色和文字不同的扇形,因此需要在 AndroidCountActivity 中创建一个 generateData()方法设置这些属性,具体代码如文件 8-5 所示。

【文件 8-5】 AndroidCountActivity.java

```
1   package com.itheima.topline.activity;
2   public class AndroidCountActivity extends AppCompatActivity {
3       private TextView tv_main_title, tv_back, tv_intro;
4       private RelativeLayout rl_title_bar;
5       private SwipeBackLayout layout;
6       private PieChartView chart;
7       private PieChartData data;
8       @Override
9       protected void onCreate(Bundle savedInstanceState) {
10          super.onCreate(savedInstanceState);
11          layout= (SwipeBackLayout) LayoutInflater.from(this).inflate(
12              R.layout.base, null);
13          layout.attachToActivity(this);
14          setContentView(R.layout.activity_android_count);
15          init();
16      }
17      private void init() {
18          tv_main_title= (TextView) findViewById(R.id.tv_main_title);
19          tv_main_title.setText("Android统计");
```

```
20      rl_title_bar=(RelativeLayout) findViewById(R.id.title_bar);
21      rl_title_bar.setBackgroundColor(getResources().getColor(R.color.
22                                                      rdTextColorPress));
23      tv_back=(TextView) findViewById(R.id.tv_back);
24      tv_back.setVisibility(View.VISIBLE);
25      tv_back.setOnClickListener(new View.OnClickListener() {
26          @Override
27          public void onClick(View view) {
28              AndroidCountActivity.this.finish();
29          }
30      });
31      tv_intro=(TextView) findViewById(R.id.tv_intro);
32      tv_intro.setText(getResources().getString(R.string.android_count_text));
33      chart=(PieChartView) findViewById(R.id.chart);
34      generateData();
35      chart.startDataAnimation();
36  }
37  private void generateData() {
38      int numValues=4;                                //设置饼状图扇形的数量
39      List<SliceValue> values=new ArrayList<SliceValue>();
40      for(int i=0; i<numValues;++i) {
41          switch(i+1) {
42              case 1:                                 //饼状图中的第一个扇形
43                  SliceValue sliceValue1=new SliceValue(i+1,
44                                          ChartUtils.COLOR_GREEN);
45                  sliceValue1.setTarget(4);           //扇形的大小
46                  sliceValue1.setLabel("月薪 8-15k"); //扇形中的文本
47                  values.add(sliceValue1);
48                  break;
49              case 2:                                 //饼状图中的第二个扇形
50                  SliceValue sliceValue2=new SliceValue(i+1,
51                                          ChartUtils.COLOR_VIOLET);
52                  sliceValue2.setTarget(3);
53                  sliceValue2.setLabel("月薪 15-20k");
54                  values.add(sliceValue2);
55                  break;
56              case 3:        //饼状图中的第三个扇形
57                  SliceValue sliceValue3=new SliceValue(i+1,
58                                          ChartUtils.COLOR_BLUE);
59                  sliceValue3.setTarget(2);
60                  sliceValue3.setLabel("月薪 20-30k");
61                  values.add(sliceValue3);
62                  break;
63              case 4:        //饼状图中的第四个扇形
64                  SliceValue sliceValue4=new SliceValue(i+1,
65                                          ChartUtils.COLOR_ORANGE);
66                  sliceValue4.setTarget(1);
67                  sliceValue4.setLabel("月薪 30k+");
68                  values.add(sliceValue4);
```

```
69                    break;
70                }
71            }
72        data=new PieChartData(values);
73        data.setHasLabels(true);
74        data.setHasLabelsOnlyForSelected(false);
75        data.setHasLabelsOutside(false);
76        chart.setPieChartData(data);
77    }
78 }
```

（4）修改 strings.xml 文件。由于"Android 统计"详情界面中的一个文本介绍信息相对较长，因此为了方便，可以把该文本写在 res/values 文件夹的 strings.xml 文件中，具体代码如下：

```
<string name="android_count_text">黑马程序员 Android+物联网课程全面升级，
以下是薪资的高低所占的比例饼状图：</string>
```

（5）修改 CountFragment.java 文件。由于点击"统计"界面的 9 个学科按钮中的第一个按钮（Android 按钮）会跳转到"Android 统计"详情界面，因此需要找到第 8 章中的文件 8-2，在该文件的 addBuilder() 方法中的注释"//跳转到 Android 统计详情界面"语句下方添加如下代码：

```
Intent android=new Intent(getActivity(), AndroidCountActivity.class);
startActivity(android);
```

（6）修改清单文件。由于"Android 统计"详情界面向右侧滑可以关闭该界面，因此需要给该界面添加透明主题的样式，在清单文件 AndroidCountActivity 对应的 activity 标签中添加如下代码：

```
android:theme="@style/AppTheme.TransparentActivity"
```

【任务 8-5】 "Java 统计"详情界面

【任务分析】

"Java 统计"详情界面主要是以柱状图和折线图的形式展示 JavaEE 工程师不同的工作经验所对应的薪资情况，界面效果如图 8-5 所示。

【任务实施】

（1）创建"Java 统计"详情界面。在 com.itheima.topline.activity 包中创建一个 Empty Activity 类，命名为 JavaCountActivity 并将布局文件名指定为 activity_java_count。

（2）放置界面控件。在布局文件中，放置一个 TextView 控件用于显示介绍信息的文本；一个 LineChartView 控件用于显示折线图；一个 ColumnChartView 控件用于显示柱状图，具体代码如文件 8-6 所示。

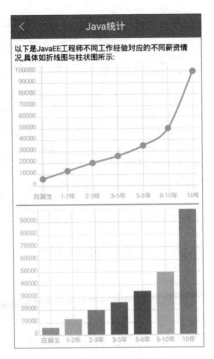

图 8-5 "Java 统计"详情界面

【文件 8-6】 activity_java_count.xml

```
1   <?xml version="1.0" encoding="utf-8"?>
2   <LinearLayout xmlns:android="http://schemas.android.com/apk/res/android"
3       android:layout_width="match_parent"
4       android:layout_height="match_parent"
5       android:background="@android:color/white"
6       android:orientation="vertical">
7       <include layout="@layout/main_title_bar" />
8       <LinearLayout
9           android:layout_width="match_parent"
10          android:layout_height="match_parent"
11          android:layout_margin="8dp"
12          android:orientation="vertical">
13          <TextView
14              android:id="@+id/tv_intro"
15              android:layout_width="match_parent"
16              android:layout_height="wrap_content"
17              android:textColor="@android:color/black"
18              android:textSize="14sp" />
19          <lecho.lib.hellocharts.view.LineChartView
20              android:id="@+id/chart_top"
21              android:layout_width="match_parent"
22              android:layout_height="0dp"
23              android:layout_marginTop="8dp"
24              android:layout_weight="1"/>
```

```xml
25    <View
26        android:layout_width="match_parent"
27        android:layout_height="2dp"
28        android:layout_marginBottom="8dp"
29        android:layout_marginTop="8dp"
30        android:background="@android:color/darker_gray" />
31    <lecho.lib.hellocharts.view.ColumnChartView
32        android:id="@+id/chart_bottom"
33        android:layout_width="match_parent"
34        android:layout_height="0dp"
35        android:layout_weight="1"/>
36  </LinearLayout>
37 </LinearLayout>
```

【任务 8-6】 "Java 统计"详情界面逻辑代码

【任务分析】

"Java 统计"详情界面主要展示了 JavaEE 工程师不同的工作经验所对应的薪资情况图,在该图中用到的柱状图与折线图分别是通过 HelloCharts 库中的 ColumnChartView 和 LineChartView 实现的。

【任务实施】

(1) 获取界面控件。在 JavaCountActivity 中创建界面控件的初始化方法 init(),用于获取"Java 统计"详情界面所要用到的控件。

(2) 设置柱状图。在 JavaCountActivity 中创建一个 generateColumnData()方法,用于设置柱状图中每个条形图的值、颜色以及横坐标的值。

(3) 设置折线图。在 JavaCountActivity 中创建 generateInitialLineData()和 generate-LineData()方法,分别用于初始化折线图与设置折线图上每个点的值,具体代码如文件 8-7 所示。

【文件 8-7】 JavaCountActivity.java

```java
1  package com.itheima.topline.activity;
2  public class JavaCountActivity extends AppCompatActivity {
3      private TextView tv_main_title, tv_back;
4      private RelativeLayout rl_title_bar;
5      private SwipeBackLayout layout;
6      public final static String[] years=new String[]{"应届生", "1-2年",
7              "2-3年", "3-5年", "5-8年", "8-10年", "10年"};
8      private LineChartView chartTop;
9      private ColumnChartView chartBottom;
10     private LineChartData lineData;
11     private ColumnChartData columnData;
12     private int[] columnY={0, 10000, 20000, 30000, 40000, 50000, 60000,
13             70000, 80000, 90000, 100000};
```

```java
14      private TextView tv_intro;
15      @Override
16      protected void onCreate(Bundle savedInstanceState) {
17          super.onCreate(savedInstanceState);
18          layout= (SwipeBackLayout) LayoutInflater.from(this).inflate(
19                  R.layout.base, null);
20          layout.attachToActivity(this);
21          setContentView(R.layout.activity_java_count);
22          init();
23      }
24      private void init() {
25          tv_main_title= (TextView) findViewById(R.id.tv_main_title);
26          tv_main_title.setText("Java统计");
27          rl_title_bar=(RelativeLayout) findViewById(R.id.title_bar);
28          rl_title_bar.setBackgroundColor(getResources().getColor(R.color.
29                  rdTextColorPress));
30          tv_intro= (TextView) findViewById(R.id.tv_intro);
31          tv_intro.setText(getResources().getString(R.string.java_count_text));
32          tv_back= (TextView) findViewById(R.id.tv_back);
33          tv_back.setVisibility(View.VISIBLE);
34          tv_back.setOnClickListener(new View.OnClickListener() {
35              @Override
36              public void onClick(View view) {
37                  JavaCountActivity.this.finish();
38              }
39          });
40          chartTop= (LineChartView) findViewById(R.id.chart_top);
41          generateInitialLineData();
42          chartBottom= (ColumnChartView) findViewById(R.id.chart_bottom);
43          generateColumnData();
44      }
45      private void generateColumnData() {
46          int numColumns=years.length;
47          List<AxisValue>axisValues=new ArrayList<AxisValue>();
48          List<AxisValue>axisYValues=new ArrayList<AxisValue>();
49          List<Column>columns=new ArrayList<Column>();
50          List<SubcolumnValue>values;
51          for(int k=0; k<columnY.length; k++) {
52              axisYValues.add(new AxisValue(k).setValue(columnY[k]));
53          }
54          for(int i=0; i<numColumns;++i) {
55              values=new ArrayList<SubcolumnValue>();
56              switch(i) {          //设置柱状图中的每个条形图的值与颜色
57                  case 0:
58                      values.add(new SubcolumnValue((float) 6000,
59                              ChartUtils.COLOR_GREEN));
60                      break;
61                  case 1:
62                      values.add(new SubcolumnValue((float) 13000,
63                              ChartUtils.COLOR_ORANGE));
```

```
64                    break;
65                case 2:
66                    values.add(new SubcolumnValue((float) 20000,
67                            ChartUtils.COLOR_BLUE));
68                    break;
69                case 3:
70                    values.add(new SubcolumnValue((float) 26000,
71                            ChartUtils.COLOR_RED));
72                    break;
73                case 4:
74                    values.add(new SubcolumnValue((float) 35000,
75                            ChartUtils.COLOR_VIOLET));
76                    break;
77                case 5:
78                    values.add(new SubcolumnValue((float) 50000,
79                            ChartUtils.COLOR_ORANGE));
80                    break;
81                case 6:
82                    values.add(new SubcolumnValue((float) 100000,
83                            ChartUtils.COLOR_BLUE));
84                    break;
85                }
86            axisValues.add(new AxisValue(i).setLabel(years[i]));
87            columns.add(new Column(values).setHasLabelsOnlyForSelected(true));
88        }
89        columnData=new ColumnChartData(columns);
90        columnData.setAxisXBottom(new Axis(axisValues).setHasLines(true));
91        columnData.setAxisYLeft(new Axis(axisYValues).setHasLines(true).
92        setMaxLabelChars(6));
93        chartBottom.setColumnChartData(columnData);
94        //这个设置会保证柱状图在点击年数的条目时条目会亮起来
95        chartBottom.setValueSelectionEnabled(true);
96        chartBottom.setZoomType(ZoomType.HORIZONTAL);
97    }
98    private void generateInitialLineData() {
99        int numValues=7;
100       List<AxisValue>axisValues=new ArrayList<AxisValue>();
101       List<PointValue>values=new ArrayList<PointValue>();
102       for(int i=0; i<numValues;++i) {
103           values.add(new PointValue(i, 0));
104           axisValues.add(new AxisValue(i).setLabel(years[i]));
105       }
106       Line line=new Line(values);
107       line.setColor(ChartUtils.COLOR_GREEN).setCubic(true);
108       List<Line>lines=new ArrayList<Line>();
109       lines.add(line);
110       lineData=new LineChartData(lines);
111       lineData.setAxisXBottom(new Axis(axisValues).setHasLines(true));
112       lineData.setAxisYLeft(new Axis().setHasLines(true).setMaxLabelChars(6));
```

```
113        chartTop.setLineChartData(lineData);
114        chartTop.setViewportCalculationEnabled(false);
115        Viewport v=new Viewport(0, 100000, 6, 0);
116        chartTop.setMaximumViewport(v);
117        chartTop.setCurrentViewport(v);
118        chartTop.setZoomType(ZoomType.HORIZONTAL);
119        generateLineData();
120    }
121    private void generateLineData() {
122        Line line=lineData.getLines().get(0);
123        for(int i=0; i<line.getValues().size(); i++) {
124            PointValue value=line.getValues().get(i);
125            switch(i) {    //设置折线图上的每个点的值
126                case 0:
127                    value.setTarget(value.getX(), (float) 6000);
128                    break;
129                case 1:
130                    value.setTarget(value.getX(), (float) 13000);
131                    break;
132                case 2:
133                    value.setTarget(value.getX(), (float) 20000);
134                    break;
135                case 3:
136                    value.setTarget(value.getX(), (float) 26000);
137                    break;
138                case 4:
139                    value.setTarget(value.getX(), (float) 35000);
140                    break;
141                case 5:
142                    value.setTarget(value.getX(), (float) 50000);
143                    break;
144                case 6:
145                    value.setTarget(value.getX(), (float) 100000);
146                    break;
147            }
148        }
149        chartTop.startDataAnimation(300);
150    }
151 }
```

（4）修改 strings.xml 文件。由于"Java 统计"详情界面中有一个文本介绍信息相对较长，因此为了方便，可以把该文本写在 res/values 文件夹中的 strings.xml 文件中，具体代码如下：

```
<string name="java_count_text">以下是 JavaEE 工程师不同的工作经验所对应的不同薪资情况，
具体如折线图与柱状图所示：</string>
```

（5）修改 CountFragment.java 文件。由于点击"统计"界面的 9 个学科按钮中的第二个按钮（Java 按钮）会跳转到"Java 统计"详情界面，因此需要在第 8 章中找到文件 8-2，在该文件的 addBuilder() 方法中的注释"//跳转到 Java 统计详情界面"下方添加如下代码：

```
Intent java=new Intent(getActivity(), JavaCountActivity.class);
startActivity(java);
```

（6）修改清单文件。由于"Java 统计"详情界面向右滑动会关闭该界面，因此需要给该界面添加透明主题的样式，在清单文件的 JavaCountActivity 对应的 activity 标签中添加如下代码：

```
android:theme="@style/AppTheme.TransparentActivity"
```

8.3 本章小结

本章主要讲解了头条项目的统计模块，统计模块主要包括"统计"界面、"Android 统计"详情界面以及"Java 统计"详情界面。"统计"界面菜单的炫酷效果与"统计详情"界面的饼状图、柱状图以及折线图的逻辑代码相对较为复杂，需要读者认真、仔细分析。

【思考题】

1. 如何实现"统计"界面菜单的炫酷效果？
2. 如何实现"统计详情"界面的饼状图、柱状图以及折线图的效果？

第 9 章 视频模块

学习目标
- 掌握"视频列表"界面的开发,实现下拉刷新功能
- 掌握"视频详情"界面的开发,实现视频播放功能

视频模块主要用于展示各学科经典视频,点击条目中的视频会进入对应的"视频"详情界面,在该界面中可以播放视频、展示视频简介与视频目录,并且可以调整播放画面的尺寸和视频清晰度等,本章将针对视频模块进行详细讲解。

9.1 视频列表

思政材料 9

任务综述

"视频列表"界面主要展示各学科的视频信息,该列表的数据从 Tomcat 服务器上获取,然后经过 JSON 解析把数据显示到"视频列表"界面上。

【任务 9-1】 "视频列表"界面

【任务分析】

"视频列表"界面主要用于展示各学科的视频信息,"视频列表"界面的效果如图 9-1 所示。

图 9-1 "视频列表"界面

【任务实施】

（1）创建"视频列表"界面。在 res/layout 文件夹中创建一个布局文件 fragment_video.xml。

（2）放置界面控件。在布局文件中，放置一个 PullToRefreshView 控件用于显示下拉刷新，一个 WrapRecyclerView 控件用于加载视频列表信息，具体代码如文件 9-1 所示。

【文件 9-1】 fragment_video.xml

```xml
<?xml version="1.0" encoding="utf-8"?>
<LinearLayout xmlns:android="http://schemas.android.com/apk/res/android"
    android:layout_width="match_parent"
    android:layout_height="match_parent"
    android:orientation="vertical">
    <com.itheima.PullToRefreshView
        android:id="@+id/pull_to_refresh"
        android:layout_width="match_parent"
        android:layout_height="match_parent"
        android:background="#f6f6f6">
        <com.itheima.topline.view.WrapRecyclerView
            android:id="@+id/recycler_view"
            android:layout_width="match_parent"
            android:layout_height="match_parent"
            android:requiresFadingEdge="none" />
    </com.itheima.PullToRefreshView>
</LinearLayout>
```

【任务 9-2】 "视频列表"界面 Item

【任务分析】

由于"视频列表"界面用到了 WrapRecyclerView 控件，因此需要为该控件创建一个 Item 界面，界面效果如图 9-2 所示。

图 9-2 "视频列表"界面 Item

【任务实施】

（1）创建"视频列表"界面 Item。在 res/layout 文件夹中创建一个布局文件 video_list_item.xml。

（2）导入界面图片。将"视频列表"界面 Item 所需的图片 media_play_icon.png 导入 drawable-hdpi 文件夹。

（3）放置界面控件。在布局文件中，放置 2 个 ImageView 控件分别用于显示视频图片和视频播放图标，具体代码如文件 9-2 所示。

【文件 9-2】 video_list_item.xml

```xml
1    <?xml version="1.0" encoding="utf-8"?>
2    <RelativeLayout
3        xmlns:android="http://schemas.android.com/apk/res/android"
4        android:layout_width="match_parent"
5        android:layout_height="150dp"
6        android:layout_marginLeft="8dp"
7        android:layout_marginRight="8dp"
8        android:layout_marginTop="8dp">
9        <ImageView
10           android:id="@+id/iv_img_round"
11           android:layout_width="fill_parent"
12           android:layout_height="fill_parent"
13           android:scaleType="fitXY" />
14       <ImageView
15           android:id="@+id/iv_media_play"
16           android:layout_width="wrap_content"
17           android:layout_height="wrap_content"
18           android:layout_centerInParent="true"
19           android:src="@drawable/media_play_icon" />
20   </RelativeLayout>
```

【任务 9-3】 创建 VideoBean

【任务分析】

由于每个学科的视频信息都有视频 Id、视频名称、视频简介、视频图片等属性，因此需要创建一个 VideoBean 类存放这些属性。

【任务实施】

（1）创建 VideoBean 类。在 com.itheima.topline.bean 包中创建一个 VideoBean 类，在该类中创建视频模块的属性，具体代码如文件 9-3 所示。

【文件 9-3】 VideoBean.java

```java
1    package com.itheima.topline.bean;
2    public class VideoBean {
3        private int id;              //视频 Id
4        private String name;         //视频名称
5        private String intro;        //视频简介
6        private String img;          //视频图片
```

```
7    private List<VideoDetailBean>videoDetailList;        //视频详情中的列表
8    public int getId() {
9        return id;
10   }
11   public void setId(int id) {
12       this.id=id;
13   }
14   public String getName() {
15       return name;
16   }
17   public void setName(String name) {
18       this.name=name;
19   }
20   public String getIntro() {
21       return intro;
22   }
23   public void setIntro(String intro) {
24       this.intro=intro;
25   }
26   public String getImg() {
27       return img;
28   }
29   public void setImg(String img) {
30       this.img=img;
31   }
32   public List<VideoDetailBean>getVideoDetailList() {
33       return videoDetailList;
34   }
35   public void setVideoDetailList(List<VideoDetailBean>videoDetailList) {
36       this.videoDetailList=videoDetailList;
37   }
38  }
```

（2）创建 VideoDetailBean 类。由于"视频详情"界面中有一个拥有自己的视频 Id 与视频名称等属性的视频目录列表，因此需要在 com.itheima.topline.bean 包中创建一个 VideoDetailBean 类，在该类中存放视频详情中视频目录的属性，具体代码如文件 9-4 所示。

【文件 9-4】 VideoDetailBean.java

```
1   package com.itheima.topline.bean;
2   public class VideoDetailBean implements Serializable {
3       private static final long serialVersionUID=1L;
4       private String video_id;
5       private String video_name;
6       public String getVideo_id() {
7           return video_id;
8       }
9       public void setVideo_id(String video_id) {
10          this.video_id=video_id;
11      }
```

```
12      public String getVideo_name() {
13          return video_name;
14      }
15      public void setVideo_name(String video_name) {
16          this.video_name=video_name;
17      }
18  }
```

【任务 9-4】 "视频列表"界面 Adapter

【任务分析】

"视频列表"界面是用 WrapRecyclerView 控件展示视频信息的,因此需要创建一个数据适配器 VideoListAdapter 对 WrapRecyclerView 控件进行数据适配。

【任务实施】

（1）创建 VideoListAdapter 类。在 com.itheima.topline.adapter 包中创建一个 VideoListAdapter 类继承 RecyclerView.Adapter<RecyclerView.ViewHolder>,并重写 onCreateViewHolder()、onBindViewHolder()、getItemCount()方法。

（2）创建 ViewHolder 类。在 VideoListAdapter 类中创建一个 ViewHolder 类以获取 Item 界面上的控件,具体代码如文件 9-5 所示。

【文件 9-5】 VideoListAdapter.java

```
1   package com.itheima.topline.adapter;
2   public class VideoListAdapter extends RecyclerView.Adapter<RecyclerView.
3   ViewHolder>{
4       private List<VideoBean>videoList;
5       private Context context;
6       public VideoListAdapter(Context context) {
7           this.context=context;
8       }
9       public void setData(List<VideoBean>videoList) {
10          this.videoList=videoList;
11          notifyDataSetChanged();
12      }
13      @Override
14      public RecyclerView.ViewHolder onCreateViewHolder(ViewGroup viewGroup,
15      int viewType) {
16          View view=LayoutInflater.from(viewGroup.getContext()).inflate(R.
17                                      layout.video_list_item, viewGroup, false);
18          ViewHolder viewHolder=new ViewHolder(view);
19          return viewHolder;
20      }
21      @Override
22      public void onBindViewHolder(final RecyclerView.ViewHolder holder, int i)
23      {
24          final VideoBean bean=videoList.get(i);
```

```
25      Glide
26              .with(context)
27              .load(bean.getImg())
28              .error(R.mipmap.ic_launcher)
29              .into(((ViewHolder) holder).iv_img);
30      holder.itemView.setOnClickListener(new View.OnClickListener() {
31          @Override
32          public void onClick(View view) {
33          }
34      });
35  }
36  @Override
37  public int getItemCount() {
38      return videoList==null ? 0 : videoList.size();
39  }
40  public class ViewHolder extends RecyclerView.ViewHolder {
41      public ImageView iv_img;
42      public ViewHolder(View itemView) {
43          super(itemView);
44          iv_img=(ImageView) itemView.findViewById(R.id.iv_img_round);
45      }
46  }
47  }
```

【任务9-5】 "视频列表"界面数据

【任务分析】

"视频列表"界面由视频图片与视频数据组成,其中,视频图片通过在Tomcat的ROOT文件夹中创建一个图片文件夹video存放,视频数据通过在ROOT文件夹中创建一个video_list_data.json文件存放。

【任务实施】

(1) 创建"视频列表"界面图片存放的文件夹。在Tomcat的ROOT/topline/img文件夹中创建一个video文件夹,用于存放视频列表界面图片。

(2) 在Tomcat服务器中创建"视频列表"界面数据文件。点击视频列表的每个条目会获取各个条目所对应的视频详情以及相关视频目录列表,由于服务器中的数据是手动添加的,如果将每个视频详情中的视频目录列表数据一个个地添加到服务器,则会产生较多的JSON文件,因此为了减少文件数量,可以在视频列表的JSON文件中存放视频详情中对应的视频目录列表信息。在Tomcat的ROOT/topline目录中创建一个video_list_data.json文件,该文件用于存放视频列表与视频详情中需要加载的数据,具体代码如文件9-6所示。

【文件9-6】 video_list_data.json

```
1   [
2     {
3       "id":1,
```

```
 4    "name":"jQuery精品教程",
 5    "img":"http://172.16.43.62:8080/topline/img/video/video_jquery_icon.png",
 6    "intro":"本视频系统讲解了jQuery的基本操作,并配合相应案例保证学生能较大程度地接受和了解
 7    知识的应用,该视频的起点都是针对有一定JavaScript基础的同学而精心设计和录制的。通过该视频
 8    的学习,相信你能够轻轻松松地掌握jQuery的应用,为学习前端开发打下坚实的基础。",
 9    "videoDetailList":[
10      {
11      "video_id":"DA2D015D371417299C33DC5901307461",
12      "video_name":"01-jQuery初体验"
13      },
14      {
15      "video_id":"8CE6F0EA79D28C409C33DC5901307461",
16      "video_name":"02-什么是jQuery"
17      },
18      {
19      "video_id":"105420027A12F7869C33DC5901307461",
20      "video_name":"03-jQuery版本问题"
21      },
22      {
23      "video_id":"2B6B824CB0FB90209C33DC5901307461",
24      "video_name":"04-jQuery入口函数的解释"
25      },
26      {
27      "video_id":"FE97256B916E64779C33DC5901307461",
28      "video_name":"05-jq对象与js对象"
29      }
30      ]
31    },
32    ......
33    {
34    "id":7,
35    "name":"MFC教程",
36    "img":"http://172.16.43.62:8080/topline/img/video/video_c_icon.png",
37    "intro":"第一天(Win消息机制、SDK编程基础),第二天(对话框、常用控件、文档和视图),
38    第三天(综合案例:销售信息管理系统)",
39    "videoDetailList":[
40      {
41      "video_id":"CC64EB602031C45E9C33DC5901307461",
42      "video_name":"01_mfc课程安排"
43      },
44      {
45      "video_id":"8002DA7F236A263E9C33DC5901307461",
46      "video_name":"02_底层窗口实现(一)_WinMain入口函数"
47      },
48      {
49      "video_id":"A8959C2D57B521A19C33DC5901307461",
50      "video_name":"03_底层窗口实现(二)_创建窗口的前五步"
51      },
52      {
53      "video_id":"F0B07D2892DE91C79C33DC5901307461",
```

```
54        "video_name":"04_底层窗口现实(三)_窗口过程处理"
55      }
56    ]
57  }
58 ]
```

（3）在 Constant 类中添加视频列表接口。由于"视频列表"界面向服务器请求数据需要一个接口地址，因此，在 com.itheima.topline.utils 包中的 Constant 类中添加如下代码：

```
//视频列表接口
public static final String REQUEST_VIDEO_URL="/video_list_data.json";
```

（4）解析 JSON 数据。由于从 Tomcat 服务器中获取的 JSON 格式的数据不能直接加载到界面上，因此需要在 com.itheima.topline.utils 包的 JsonParse 类中创建一个 getVideoList()方法，用于解析该界面获取的 JSON 数据，需要在 JsonParse 类中添加如下代码：

```
1   public List<VideoBean>getVideoList(String json) {
2       //使用 gson 库解析 JSON 数据
3       Gson gson=new Gson();
4       //创建一个 TypeToken 的匿名子类对象,并调用对象的 getType()方法
5       Type listType=new TypeToken<List<VideoBean>>() {
6       }.getType();
7       //把获取的信息集合存入 videoList
8       List<VideoBean>videoList=gson.fromJson(json, listType);
9       return videoList;
10  }
```

【任务 9-6】"视频列表"界面逻辑代码

【任务分析】

"视频列表"界面主要是一个展示视频信息的列表，该界面中的数据是通过解析 Tomcat 中的 video_list_data.json 文件而得到的，然后把解析后的数据加载到"视频列表"界面。

【任务实施】

（1）创建 VideoFragment 类。在 com.itheima.topline.fragment 包中创建一个 VideoFragment 类。在该类中创建界面控件的初始化方法 initView()，在该方法中获取"视频列表"界面需要用到的 UI 控件并初始化。

（2）从服务器获取数据。在 VideoFragment 中创建一个 initData()方法，用于从 Tomcat 服务器中获取视频列表数据，具体代码如文件 9-7 所示。

【文件 9-7】 VideoFragment.java

```
1   package com.itheima.topline.fragment;
2   public class VideoFragment extends Fragment {
```

```java
3      private PullToRefreshView mPullToRefreshView;
4      private WrapRecyclerView recyclerView;
5      public static final int REFRESH_DELAY=1000;
6      public static final int MSG_VIDEO_OK=1;           //获取数据
7      private MHandler mHandler;                         //事件捕获
8      private VideoListAdapter adapter;
9      public VideoFragment() {
10     }
11     @Override
12     public View onCreateView(LayoutInflater inflater, ViewGroup container,
13     Bundle savedInstanceState) {
14         mHandler=new MHandler();
15         initData();
16         View view=initView(inflater, container);
17         return view;
18     }
19     private View initView(LayoutInflater inflater, ViewGroup container) {
20         View view=inflater.inflate(R.layout.fragment_video, container, false);
21         recyclerView= (WrapRecyclerView) view.findViewById(R.id.recycler_view);
22         recyclerView.setLayoutManager(new LinearLayoutManager(getContext()));
23         adapter=new VideoListAdapter(getActivity());
24         recyclerView.setAdapter(adapter);
25         mPullToRefreshView= (PullToRefreshView) view.findViewById(R.id.
26         pull_to_refresh);
27         mPullToRefreshView.setOnRefreshListener(new PullToRefreshView.
28         OnRefreshListener() {
29             @Override
30             public void onRefresh() {
31                 mPullToRefreshView.postDelayed(new Runnable() {
32                     @Override
33                     public void run() {
34                         mPullToRefreshView.setRefreshing(false);
35                         initData();
36                     }
37                 }, REFRESH_DELAY);
38             }
39         });
40         return view;
41     }
42     private void initData() {
43         OkHttpClient okHttpClient=new OkHttpClient();
44         Request request=new Request.Builder().url(Constant.WEB_SITE+
45                         Constant.REQUEST_VIDEO_URL).build();
46         Call call=okHttpClient.newCall(request);
47         //开启异步线程访问网络
48         call.enqueue(new Callback() {
49             @Override
50             public void onResponse(Response response) throws IOException {
51                 String res=response.body().string();
52                 Message msg=new Message();
```

```
53                msg.what=MSG_VIDEO_OK;
54                msg.obj=res;
55                mHandler.sendMessage(msg);
56            }
57            @Override
58            public void onFailure(Request arg0, IOException arg1) {
59            }
60        });
61    }
62    /**
63     * 事件捕获
64     */
65    class MHandler extends Handler {
66        @Override
67        public void dispatchMessage(Message msg) {
68            super.dispatchMessage(msg);
69            switch(msg.what) {
70                case MSG_VIDEO_OK:
71                    if(msg.obj !=null) {
72                        String vlResult= (String) msg.obj;
73                        //使用 gson 解析数据
74                        List<VideoBean>videoList=JsonParse.getInstance().
75                        getVideoList(vlResult);
76                        adapter.setData(videoList);
77                    }
78                    break;
79            }
80        }
81    }
82 }
```

（3）修改底部导航栏。由于点击底部导航栏的"视频"按钮时会出现"视频列表"界面，因此需要找到第 6 章文件 6-8 中的 initView（）方法，在该方法的"CountFragment countFragment＝new CountFragment（）;"语句下方添加如下代码：

```
VideoFragment videoFragment=new VideoFragment();
```

在"alFragment.add(countFragment);"语句下方添加如下代码：

```
alFragment.add(videoFragment);
```

9.2 视频详情

任务综述

"视频详情"界面主要显示视频简介、视频目录、视频播放等信息，由于该界面的数据是从"视频列表"界面传递过来的，因此只需要把传递过来的数据展示到界面即可。"视频详情"界面中的视频使用第三方 CC 视频播放器播放。

【任务 9-7】 "视频详情"界面

【任务分析】

点击视频列表的条目会跳转到"视频详情"界面,该界面主要显示视频简介、视频目录以及视频播放等信息,界面效果如图 9-3 所示。

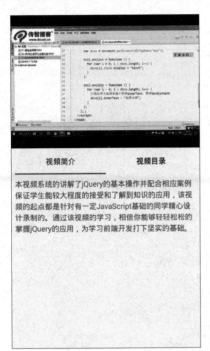

图 9-3 "视频详情"界面

【任务实施】

(1) 创建"视频详情"界面。在 com.itheima.topline.activity 包中创建一个 Empty Activity 类,命名为 VideoDetailActivity,并将布局文件名指定为 activity_video_detail。

(2) 导入界面图片。将"视频详情"界面所需的图片 big_stop_ic.png、back.png、more_ic.png、up_ic.png、smallbegin_ic.png、smallstop_ic.png、down_ic.png、fullscreen_open.png、fullscreen_close.png、volume.png、lock_ic.png、unlock_ic.png、popdown.9.png 导入 drawable-hdpi 文件夹。

(3) 添加 design 库。由于"视频详情"界面用到了 design 库中的 TabLayout 类,因此需要在该项目中添加 design 库。在 Android Studio 中,选中项目,右击选择 Open Module Settings 选项后选择 Dependencies 选项卡,单击右上角的绿色加号,选择 Library dependency 选项,把 com.android.support:design:25.3.1 库加入主项目。

(4) 放置界面控件。该布局文件中的控件相对较多,读者若不想一一编写,则可以直接从源代码中复制,然后梳理布局结构,具体代码如文件 9-8 所示。

【文件9-8】 activity_video_detail.xml

```xml
1   <?xml version="1.0" encoding="utf-8"?>
2   <LinearLayout xmlns:android="http://schemas.android.com/apk/res/android"
3       xmlns:app="http://schemas.android.com/apk/res-auto"
4       android:layout_width="fill_parent"
5       android:layout_height="fill_parent"
6       android:orientation="vertical">
7       <RelativeLayout
8           android:id="@+id/rl_play"
9           android:layout_width="fill_parent"
10          android:layout_height="0dp"
11          android:layout_weight="2"
12          android:background="#000000">
13          <SurfaceView
14              android:id="@+id/playerSurfaceView"
15              android:layout_width="wrap_content"
16              android:layout_height="wrap_content"
17              android:layout_centerInParent="true"/>
18          <ProgressBar
19              android:id="@+id/bufferProgressBar"
20              style="?android:attr/progressBarStyleLarge"
21              android:layout_width="wrap_content"
22              android:layout_height="wrap_content"
23              android:layout_centerInParent="true" />
24          <ImageView
25              android:id="@+id/iv_center_play"
26              android:layout_width="60dp"
27              android:layout_height="60dp"
28              android:layout_centerInParent="true"
29              android:src="@drawable/big_stop_ic"
30              android:visibility="gone" />
31          <LinearLayout
32              android:id="@+id/playerTopLayout"
33              android:layout_width="fill_parent"
34              android:layout_height="45dp"
35              android:layout_alignParentTop="true"
36              android:layout_gravity="top"
37              android:background="@drawable/play_top_bg"
38              android:orientation="horizontal"
39              android:padding="3dp"
40              android:visibility="gone">
41              <ImageView
42                  android:id="@+id/backPlayList"
43                  android:layout_width="wrap_content"
44                  android:layout_height="wrap_content"
45                  android:layout_gravity="center_vertical"
46                  android:src="@drawable/back" />
47              <TextView
48                  android:id="@+id/videoIdText"
```

```xml
49              android:layout_width="0dp"
50              android:layout_height="wrap_content"
51              android:layout_gravity="center_vertical"
52              android:layout_weight="1"
53              android:gravity="left"
54              android:singleLine="true"
55              android:textColor="#FFFFFFFF"
56              android:textSize="16sp" />
57          <ImageView
58              android:id="@+id/iv_top_menu"
59              android:layout_width="45dp"
60              android:layout_height="45dp"
61              android:layout_gravity="center_vertical"
62              android:layout_marginLeft="15dp"
63              android:layout_marginRight="20dp"
64              android:padding="5dp"
65              android:scaleType="fitXY"
66              android:src="@drawable/more_ic" />
67      </LinearLayout>
68      <LinearLayout
69          android:id="@+id/playerBottomLayout"
70          android:layout_width="fill_parent"
71          android:layout_height="50dp"
72          android:layout_alignParentBottom="true"
73          android:background="#B2000000"
74          android:orientation="horizontal"
75          android:visibility="gone">
76          <LinearLayout
77              android:layout_width="wrap_content"
78              android:layout_height="wrap_content"
79              android:layout_gravity="center_vertical"
80              android:gravity="center"
81              android:orientation="horizontal">
82              <ImageView
83                  android:id="@+id/iv_video_back"
84                  android:layout_width="30dp"
85                  android:layout_height="wrap_content"
86                  android:layout_marginRight="5dp"
87                  android:src="@drawable/up_ic" />
88              <ImageView
89                  android:id="@+id/iv_play"
90                  android:layout_width="30dp"
91                  android:layout_height="wrap_content"
92                  android:src="@drawable/smallstop_ic" />
93              <ImageView
94                  android:id="@+id/iv_video_next"
95                  android:layout_width="30dp"
96                  android:layout_height="wrap_content"
97                  android:layout_marginLeft="5dp"
98                  android:src="@drawable/down_ic" />
```

```xml
 99            </LinearLayout>
100            <TextView
101                android:id="@+id/playDuration"
102                android:layout_width="wrap_content"
103                android:layout_height="wrap_content"
104                android:layout_gravity="center_vertical"
105                android:layout_marginLeft="5dp"
106                android:textColor="#FFFFFF" />
107            <SeekBar
108                android:id="@+id/skbProgress"
109                android:layout_width="0dp"
110                android:layout_height="wrap_content"
111                android:layout_gravity="center_vertical"
112                android:layout_marginLeft="10dp"
113                android:layout_weight="1"
114                android:maxHeight="3dp"
115                android:minHeight="3dp"
116                android:progressDrawable="@drawable/seekbar_style"/>
117            <TextView
118                android:id="@+id/videoDuration"
119                android:layout_width="wrap_content"
120                android:layout_height="wrap_content"
121                android:layout_gravity="center_vertical"
122                android:layout_marginRight="8dp"
123                android:textColor="#FFFFFF" />
124            <ImageView
125                android:id="@+id/iv_fullscreen"
126                android:layout_width="40dp"
127                android:layout_height="40dp"
128                android:layout_gravity="center_vertical"
129                android:scaleType="centerInside"
130                android:src="@drawable/fullscreen_close" />
131            <LinearLayout
132                android:layout_width="wrap_content"
133                android:layout_height="wrap_content"
134                android:layout_gravity="center_vertical"
135                android:gravity="center"
136                android:orientation="horizontal">
137                <!--倍速播放选择-->
138                <TextView
139                    android:id="@+id/tv_speed_play"
140                    style="@style/playBottomTextViewStyle"
141                    android:text="@string/speed" />
142                <TextView
143                    android:id="@+id/tv_definition"
144                    style="@style/playBottomTextViewStyle"
145                    android:text="@string/definition" />
146            </LinearLayout>
147        </LinearLayout>
148        <LinearLayout
```

```xml
149            android:id="@+id/volumeLayout"
150            android:layout_width="30dp"
151            android:layout_height="wrap_content"
152            android:layout_alignParentRight="true"
153            android:layout_centerVertical="true"
154            android:layout_marginRight="30dp"
155            android:background="#80000000"
156            android:gravity="center_horizontal"
157            android:orientation="vertical"
158            android:visibility="gone">
159            <com.itheima.topline.view.VerticalSeekBar
160                android:id="@+id/volumeSeekBar"
161                android:layout_width="wrap_content"
162                android:layout_height="200dp"
163                android:maxHeight="5dp"
164                android:minHeight="5dp"
165                android:progressDrawable="@drawable/seekbar_style" />
166            <ImageView
167                android:layout_width="20dp"
168                android:layout_height="20dp"
169                android:layout_marginBottom="10dp"
170                android:src="@drawable/volume" />
171        </LinearLayout>
172        <ImageView
173            android:id="@+id/iv_lock"
174            android:layout_width="50dp"
175            android:layout_height="50dp"
176            android:layout_centerVertical="true"
177            android:layout_marginLeft="10dp"
178            android:padding="5dp"
179            android:scaleType="fitCenter"
180            android:src="@drawable/player_lock_bg"
181            android:visibility="gone" />
182    </RelativeLayout>
183    <LinearLayout
184        android:id="@+id/ll_below_info"
185        android:layout_width="fill_parent"
186        android:layout_height="0dp"
187        android:layout_weight="3"
188        android:orientation="vertical">
189        <android.support.design.widget.TabLayout
190            android:id="@+id/tabs"
191            android:layout_width="match_parent"
192            android:layout_height="wrap_content"
193            android:layout_marginLeft="8dp"
194            android:layout_marginRight="8dp"
195            app:tabIndicatorColor="@android:color/holo_red_dark"
196            app:tabSelectedTextColor="@android:color/holo_red_dark"
197            app:tabTextColor="@android:color/black" />
198        <!--可滑动的布局内容-->
```

```
199        <android.support.v4.view.ViewPager
200            android:id="@+id/vp_view"
201            android:layout_width="match_parent"
202            android:layout_height="wrap_content" />
203    </LinearLayout>
204 </LinearLayout>
```

(5) 创建 play_top_bg.xml 文件。由于"视频详情"界面需要一个渐变背景,因此需要在 res/drawable 文件夹中创建一个 play_top_bg.xml 文件,具体代码如文件 9-9 所示。

【文件 9-9】 play_top_bg.xml

```
<?xml version="1.0" encoding="utf-8"?>
<shape xmlns:android="http://schemas.android.com/apk/res/android">
    <gradient android:startColor="#CC000000" android:endColor="#00000000"
        android:angle="270"/>
</shape>
```

上述代码中,shape 用于定义形状,gradient 用于定义该形状为渐变色填充,startColor 为起始颜色,endColor 为结束颜色,angle 表示渐变角度,且必须是 45 的整数倍。

(6) 修改 strings.xml 文件。在 res/values 文件夹的 strings.xml 文件中添加如下代码:

```
<string name="definition">清晰度</string>
<string name="speed">倍速</string>
```

(7) 创建播放条 SeekBar 的样式。由于播放视频时的播放条具有一定的样式,因此需要在 res/drawable 文件夹中创建一个 seekbar_style.xml 文件,用于设置播放条的样式,具体代码如文件 9-10 所示。

【文件 9-10】 seekbar_style.xml

```
1  <layer-list xmlns:android="http://schemas.android.com/apk/res/android">
2      <item android:id="@android:id/background">
3          <shape>
4              <corners android:radius="10dip" />
5              <gradient
6                  android:angle="270"
7                  android:centerColor="#151515"
8                  android:centerY="0.45"
9                  android:endColor="#151515"
10                 android:startColor="#151515" />
11         </shape>
12     </item>
13     <item android:id="@android:id/secondaryProgress">
14         <clip>
15             <shape>
16                 <corners android:radius="10dip" />
17                 <gradient
18                     android:angle="270"
```

```
19                 android:centerColor="#333333"
20                 android:centerY="0.45"
21                 android:endColor="#333333"
22                 android:startColor="#333333" />
23            </shape>
24        </clip>
25    </item>
26    <item android:id="@android:id/progress">
27        <clip>
28            <shape>
29                <corners android:radius="10dip" />
30                <gradient
31                    android:angle="270"
32                    android:centerColor="#a07e5d"
33                    android:centerY="0.45"
34                    android:endColor="#a07e5d"
35                    android:startColor="#a07e5d" />
36            </shape>
37        </clip>
38    </item>
39 </layer-list>
```

（8）创建清晰度播放选择的文本样式。由于在横屏播放视频时有选择清晰度的文本，这些文本都使用相同的样式，因此需要在 res/values 文件夹的 styles.xml 文件中添加名为 playBottomTextViewStyle 的样式，具体代码如下所示：

```
<style name="playBottomTextViewStyle" parent="android:Widget.Holo.Light.TextView">
    <item name="android:layout_marginRight">5dp</item>
    <item name="android:textColor">#FFFFFF</item>
    <item name="android:textSize">13sp</item>
    <item name="android:visibility">gone</item>
    <item name="android:padding">5dp</item>
    <item name="android:layout_width">wrap_content</item>
    <item name="android:layout_height">wrap_content</item>
    <item name="android:layout_gravity">center_vertical</item>
</style>
```

（9）创建视频的锁屏图标。由于在横屏播放时有锁屏图标，当点击锁屏时，会显示一个上锁的图标（lock_ic.png），当再次点击开锁时，会显示一个开锁的图标（unlock_ic.png），因此为了实现这个效果，需要在 res/drawable 文件夹中创建一个 player_lock_bg.xml 文件，具体代码如文件 9-11 所示。

【文件 9-11】 player_lock_bg.xml

```
1 <?xml version="1.0" encoding="utf-8"?>
2 <selector xmlns:android="http://schemas.android.com/apk/res/android">
3     <item android:drawable="@drawable/lock_ic" android:state_selected="true" />
4     <item android:drawable="@drawable/unlock_ic" android:state_selected="false" />
5 </selector>
```

（10）创建视频简介布局和视频目录布局。由于"视频详情"界面还包含一个视频简介与一个视频目录，因此需要在 res/layout 文件夹中创建 video_detail_viewpager1.xml 文件与 video_detail_viewpager2.xml 文件，分别用于显示视频简介布局与视频目录布局，具体代码如文件 9-12 和文件 9-13 所示。

【文件 9-12】 video_detail_viewpager1.xml

```xml
1  <?xml version="1.0" encoding="utf-8"?>
2  <LinearLayout xmlns:android="http://schemas.android.com/apk/res/android"
3      android:layout_width="match_parent"
4      android:layout_height="match_parent">
5      <TextView
6          android:id="@+id/tv_intro"
7          android:layout_width="match_parent"
8          android:layout_height="match_parent"
9          android:layout_marginTop="8dp"
10         android:lineSpacingExtra="4dp"
11         android:paddingLeft="8dp"
12         android:paddingRight="8dp" />
13 </LinearLayout>
```

【文件 9-13】 video_detail_viewpager2.xml

```xml
1  <?xml version="1.0" encoding="utf-8"?>
2  <LinearLayout xmlns:android="http://schemas.android.com/apk/res/android"
3      android:layout_width="match_parent"
4      android:layout_height="match_parent"
5      android:orientation="vertical">
6      <View
7          android:layout_width="fill_parent"
8          android:layout_height="1dp"
9          android:background="#e1e1e1" />
10     <ListView
11         android:id="@+id/lv_list"
12         android:layout_width="match_parent"
13         android:layout_height="match_parent" />
14 </LinearLayout>
```

【任务 9-8】 "视频目录"列表 Item

【任务分析】

由于"视频详情"界面的视频目录用到了 ListView 控件，因此需要为该控件创建一个 Item 布局，界面效果如图 9-4 所示。

图 9-4 "视频目录"列表的 Item

【任务实施】

（1）创建"视频目录"列表 Item。在 res/layout 文件夹中创建一个布局文件 video_detail_item.xml。

（2）导入界面图片。将"视频列表"界面 Item 所需的图片 iv_video_icon.png、iv_video_selected_icon.png 导入 drawable-hdpi 文件夹。

（3）放置界面控件。在布局文件中，放置一个 TextView 控件用于显示视频名称；一个 ImageView 控件用于显示播放图标，具体代码如文件 9-14 所示。

【文件 9-14】 video_detail_item.xml

```
1  <?xml version="1.0" encoding="utf-8"?>
2  <RelativeLayout xmlns:android="http://schemas.android.com/apk/res/android"
3      android:layout_width="match_parent"
4      android:layout_height="match_parent"
5      android:padding="10dp">
6      <TextView
7          android:id="@+id/tv_video_name"
8          android:layout_width="wrap_content"
9          android:layout_height="wrap_content"
10         android:layout_alignParentLeft="true"
11         android:layout_centerVertical="true" />
12     <ImageView
13         android:id="@+id/iv_icon"
14         android:layout_width="25dp"
15         android:layout_height="25dp"
16         android:layout_alignParentRight="true"
17         android:layout_centerVertical="true"
18         android:scaleType="fitXY"
19         android:src="@drawable/iv_video_icon" />
20 </RelativeLayout>
```

【任务 9-9】 画面尺寸菜单

【任务分析】

在"视频详情"界面点击正在播放的视频右下角的"全屏"图标，界面会变成横屏显示，当点击"视频"界面右上角的"更多"按钮时，会弹出画面尺寸选择的菜单，界面效果如图 9-5 所示。

【任务实施】

（1）创建"画面尺寸菜单"界面。在 res/layout 文件夹中创建一个布局文件 play_top_menu.xml。

（2）放置界面控件。在布局文件中，放置一个 TextView 控件用于显示"画面尺寸"文字；4 个 RadioButton 控件分别用于显示满屏、100％、75％以及 50％四个按钮，具体代码如文件 9-15 所示。

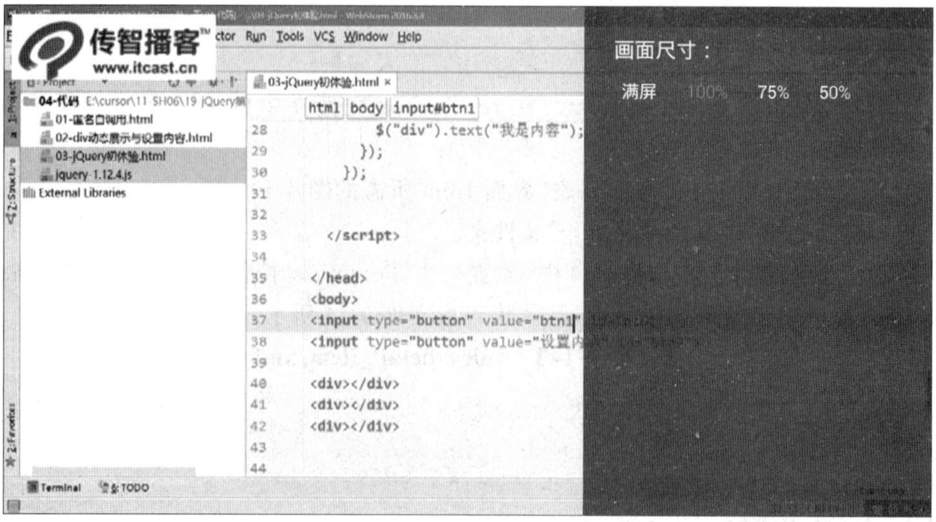

图 9-5 画面尺寸菜单

【文件 9-15】 play_top_menu.xml

```
1   <?xml version="1.0" encoding="utf-8"?>
2   <LinearLayout xmlns:android="http://schemas.android.com/apk/res/android"
3       android:layout_width="match_parent"
4       android:layout_height="match_parent"
5       android:orientation="vertical"
6       android:paddingLeft="15dp"
7       android:paddingTop="5dp">
8       <TextView
9           android:layout_width="wrap_content"
10          android:layout_height="wrap_content"
11          android:layout_gravity="center_vertical"
12          android:layout_marginLeft="5dp"
13          android:layout_marginTop="15dp"
14          android:text="画面尺寸:"
15          android:textColor="#FFFFFF"
16          android:textSize="14sp" />
17      <RadioGroup
18          android:id="@+id/rg_screensize"
19          android:layout_width="wrap_content"
20          android:layout_height="wrap_content"
21          android:layout_margin="5dp"
22          android:orientation="horizontal">
23          <RadioButton
24              android:id="@+id/rb_screensize_full"
25              style="@style/rbStyle"
26              android:text="满屏" />
27          <RadioButton
28              android:id="@+id/rb_screensize_100"
29              style="@style/rbStyle"
30              android:layout_marginLeft="10dp"
```

```
31                    android:checked="true"
32                    android:text="100% " />
33              <RadioButton
34                    android:id="@+id/rb_screensize_75"
35                    style="@style/rbStyle"
36                    android:layout_marginLeft="10dp"
37                    android:text="75% " />
38              <RadioButton
39                    android:id="@+id/rb_screensize_50"
40                    style="@style/rbStyle"
41                    android:layout_marginLeft="10dp"
42                    android:text="50% " />
43        </RadioGroup>
44 </LinearLayout>
```

(3) 修改 styles.xml 文件。由于在画"面尺寸菜单"界面中的满屏、100%、75%以及 50%这 4 个按钮的样式是相同的,因此需要在 res/values 文件夹的 styles.xml 文件中添加一个名为 rbStyle 的样式,具体代码如下所示:

```
<style name="rbStyle">
    <item name="android:button">@null</item>
    <item name="android:padding">5dp</item>
    <item name="android:textColor">@drawable/play_rb_textcolor</item>
    <item name="android:textSize">12sp</item>
</style>
```

(4) 创建 play_rb_textcolor.xml 文件。由于画"面尺寸菜单"界面中的 4 个按钮在点击后字体颜色会发生变化,以"满屏"按钮为例,当未点击"满屏"按钮时,按钮上的文字为白色;当点击"满屏"按钮时,按钮上的文字为橙色,因此需要在 res/drawable 文件夹中创建一个 play_rb_textcolor.xml 文件(文字颜色选择器),具体代码如文件 9-16 所示。

【文件 9-16】 play_rb_textcolor.xml

```
1  <?xml version="1.0" encoding="utf-8"?>
2  <selector xmlns:android="http://schemas.android.com/apk/res/android">
3      <item android:state_checked="true" android:color="@color/rb_text_check" />
4      <item android:color="@android:color/white" android:state_checked="false" />
5  </selector>
```

(5) 修改 colors.xml 文件。由于 play_rb_textcolor.xml 文件用到了 rb_text_check 颜色值,因此需要在 res/values 文件夹的 colors.xml 文件中添加如下代码:

```
<color name="rb_text_check">#ff5200</color>
```

【任务 9-10】 "视频目录"列表 Adapter

【任务分析】

"视频目录"列表是通过 ListView 控件展示视频目录信息的,因此需要创建一个数据适配器 VideoDetailListAdapter 对 ListView 控件进行数据适配。

【任务实施】

（1）创建 VideoDetailListAdapter 类。在 com.itheima.topline.adapter 包中创建一个 VideoDetailListAdapter 类继承 BaseAdapter 类，并重写 getCount()、getItem()、getItemId()、getView() 方法。在 getView() 方法中设置 XML 布局、数据以及界面跳转方法，同时，为了减少缓存，需要复用 convertView 对象，具体代码如文件 9-17 所示。

【文件 9-17】 VideoDetailListAdapter.java

```
1  package com.itheima.topline.adapter;
2  public class VideoDetailListAdapter extends BaseAdapter {
3      private Context mContext;
4      private List<VideoDetailBean>vdbl;
5      private int selectedPosition=-1;          //点击时选中的位置
6      private OnSelectListener onSelectListener;
7      public VideoDetailListAdapter(Context context, OnSelectListener
8      onSelectListener) {
9          this.mContext=context;
10         this.onSelectListener=onSelectListener;
11     }
12     public void setSelectedPosition(int position) {
13         selectedPosition=position;
14     }
15     /**
16      * 设置数据,更新界面
17      */
18     public void setData(List<VideoDetailBean>vdbl) {
19         this.vdbl=vdbl;
20         notifyDataSetChanged();
21     }
22     /**
23      * 获取 Item 的总数
24      */
25     @Override
26     public int getCount() {
27         return vdbl==null ? 0 : vdbl.size();
28     }
29     /**
30      * 根据 position 得到对应 Item 的对象
31      */
32     @Override
33     public VideoDetailBean getItem(int position) {
34         return vdbl==null ? null : vdbl.get(position);
35     }
36     /**
37      * 根据 position 得到对应 Item 的 Id
38      */
39     @Override
```

```java
40      public long getItemId(int position) {
41          return position;
42      }
43      /**
44       * 得到相应 position 对应的 Item 视图,参数 position 是当前 Item 的位置
45       * 参数 convertView 就是滑出屏幕的 Item 的 View
46       */
47      @Override
48      public View getView(final int position, View convertView, ViewGroup parent)
49      {
50          final ViewHolder vh;
51          //复用 convertView
52          if(convertView==null) {
53              vh=new ViewHolder();
54              convertView=LayoutInflater.from(mContext).inflate(
55              R.layout.video_detail_item, null);
56              vh.title=(TextView) convertView.findViewById(R.id.tv_video_name);
57              vh.iv_icon= (ImageView) convertView.findViewById(R.id.iv_icon);
58              convertView.setTag(vh);
59          } else {
60              vh= (ViewHolder) convertView.getTag();
61          }
62          //获取 position 对应的 Item 的数据对象
63          final VideoDetailBean bean=getItem(position);
64          vh.title.setTextColor(mContext.getResources().getColor(R.color.
65          video_detail_text_color));
66          if(bean !=null) {
67              vh.title.setText(bean.getVideo_name());
68              //设置选中效果
69              if(selectedPosition==position) {
70                  vh.iv_icon.setImageResource(R.drawable.iv_video_selected_icon);
71                  vh.title.setTextColor(mContext.getResources().getColor(
72                              R.color.rdTextColorPress));
73              } else {
74                  vh.iv_icon.setImageResource(R.drawable.iv_video_icon);
75                  vh.title.setTextColor(mContext.getResources().getColor(
76                              R.color.video_detail_text_color));
77              }
78          }
79          //每个 Item 的点击事件
80          convertView.setOnClickListener(new View.OnClickListener() {
81              @Override
82              public void onClick(View v) {
83                  //跳转到习题详情界面
84                  if(bean==null) {
85                      return;
86                  }
87                  //播放视频
88                  onSelectListener.onSelect(position, vh.iv_icon);
89              }
```

```
 90              });
 91              return convertView;
 92          }
 93      class ViewHolder {
 94          public TextView title;
 95          public ImageView iv_icon;
 96      }
 97      /**
 98       * 创建 OnSelectListener 接口,把位置 position 和控件 ImageView 传递到 Activity 界面
 99       */
100      public interface OnSelectListener {
101          void onSelect(int position, ImageView iv);
102      }
103  }
```

(2)修改 colors.xml 文件。由于视频目录标题多次使用同一个颜色值,因此为了减少代码量与方便调用,需要把视频目录文本的颜色值添加到 res/values 文件夹的 colors.xml 文件中,添加的代码如下:

```
<color name="video_detail_text_color">#333333</color>
```

【任务 9-11】 创建 TopLineApplication

【任务分析】

由于"视频播放"界面用到了第三方 CC 视频播放器,同时在播放视频之前需要启动服务,因此在该项目中需要创建一个 Application 用于处理 CC 视频播放器用到的服务。

【任务实施】

(1)添加 CCSDK.jar 库。由于视频播放调用的是第三方 CC 视频播放器,因此需要在 Project 选项卡下将 CCSDK.jar 复制到 app 中的 libs 文件夹,选中 CCSDK.jar 库,右击选择 Add As Library 选项,会弹出一个对话框,把该 jar 包放在 app 项目中即可。

(2)创建 TopLineApplication.java 文件。由于 CC 视频播放器在播放视频时需要开启 DRMServer 服务,因此需要在 com.itheima.topline 文件夹中创建一个 TopLineApplication 类,用于设置视频播放时用到的服务,具体代码如文件 9-18 所示。

【文件 9-18】 TopLineApplication.java

```
 1  package com.itheima.topline;
 2  public class TopLineApplication extends Application {
 3      private DRMServer drmServer;          //视频播放时用到的服务
 4      @Override
 5      public void onCreate() {
 6          startDRMServer();
 7          super.onCreate();
 8      }
 9      //启动 DRMServer
10      public void startDRMServer() {
```

```
11        if(drmServer==null) {
12            drmServer=new DRMServer();
13            drmServer.setRequestRetryCount(10);
14        }
15        try {
16            drmServer.start();
17            setDrmServerPort(drmServer.getPort());
18        } catch(Exception e) {
19            Toast.makeText(getApplicationContext(),"启动解密服务失败,请检查网络限制情况",
20                Toast.LENGTH_LONG).show();
21        }
22    }
23    @Override
24    public void onTerminate() {
25        if(drmServer !=null) {
26            drmServer.stop();
27        }
28        super.onTerminate();
29    }
30    private int drmServerPort;
31    public int getDrmServerPort() {
32        return drmServerPort;
33    }
34    public void setDrmServerPort(int drmServerPort) {
35        this.drmServerPort=drmServerPort;
36    }
37    public DRMServer getDRMServer() {
38        return drmServer;
39    }
40 }
```

（3）修改清单文件。在 AndroidManifest.xml 文件中需要设置 application 标签中的 name 属性值为".TopLineApplication"。

【任务 9-12】 创建 VideoDetailPagerAdapter

【任务分析】

由于"视频详情"界面用到了 ViewPager 控件,因此需要创建一个 VideoDetailPagerAdapter 对 ViewPager 控件进行数据填充。

【任务实施】

在 com.itheima.topline.adapter 文件夹中创建一个 VideoDetailPagerAdapter 类并继承 PagerAdapter 类,具体代码如文件 9-19 所示。

【文件 9-19】 VideoDetailPagerAdapter.java

```
1  package com.itheima.topline.adapter;
2  public class VideoDetailPagerAdapter extends PagerAdapter {
```

```
3      private List<View>mViewList;
4      private List<String>mTitleList;
5      public VideoDetailPagerAdapter(List<View>mViewList, List<String>mTitleList){
6          this.mViewList=mViewList;
7          this.mTitleList=mTitleList;
8      }
9      @Override
10     public int getCount() {
11         return mViewList.size();           //页卡数
12     }
13     @Override
14     public boolean isViewFromObject(View view, Object object) {
15         return view==object;               //官方推荐写法
16     }
17     @Override
18     public Object instantiateItem(ViewGroup container, int position) {
19         container.addView(mViewList.get(position));     //添加页卡
20         return mViewList.get(position);
21     }
22     @Override
23     public void destroyItem(ViewGroup container, int position, Object object) {
24         container.removeView(mViewList.get(position));  //删除页卡
25     }
26     @Override
27     public CharSequence getPageTitle(int position) {
28         return mTitleList.get(position);    //页卡标题
29     }
30 }
```

【任务 9-13】 创建 ParamsUtils

【任务分析】

由于在"视频详情"界面中需要把时间转换成字符串显示到界面上,并且还需要把 dp 转换成 px,因此需要创建一个工具类,并在该类中分别创建两个方法实现这些功能。

【任务实施】

在 com.itheima.topline.utils 包中创建一个 ParamsUtils 类,在该类中创建 millsecondsToStr() 方法与 dpToPx() 方法,分别用于把时间转换成字符串,以及把 dp 转换成 px,具体代码如文件 9-20 所示。

【文件 9-20】 ParamsUtils.java

```
1  package com.itheima.topline.utils;
2  public class ParamsUtils {
3      public final static int INVALID=-1;
4      public static int getInt(String str){
5          int num=INVALID;
6          try {
```

```
7         num=Integer.parseInt(str);
8     } catch(NumberFormatException e) {
9     }
10    return num;
11 }
12 public static String millsecondsToStr(int seconds){    //把时间转换成字符串
13    seconds=seconds / 1000;
14    String result="";
15    int hour=0, min=0, second=0;
16    hour=seconds / 3600;
17    min=(seconds - hour * 3600) / 60;
18    second=seconds - hour * 3600 - min * 60;
19    if(hour<10) {
20        result+="0"+hour+":";
21    } else {
22        result+=hour+":";
23    }
24    if(min<10) {
25        result+="0"+min+":";
26    } else {
27        result+=min+":";
28    }
29    if(second<10) {
30        result+="0"+second;
31    } else {
32        result+=second;
33    }
34    return result;
35 }
36 public static int dpToPx(Context context, int height){    //dp 转换成 px
37    float density=context.getResources().getDisplayMetrics().density;
38    height=(int) (height * density+0.5f);
39    return height;
40 }
41 }
```

【任务 9-14】 视频播放进度条

【任务分析】

由于在"视频详情"界面播放视频时会显示视频播放的进度,因此需要创建一个视频播放的进度条。

【任务实施】

在 com.itheima.topline.view 文件夹中创建一个 VerticalSeekBar 类用于设置视频播放时的进度条,具体代码如文件 9-21 所示。

【文件 9-21】 VerticalSeekBar.java

```java
1  package com.itheima.topline.view;
2  public class VerticalSeekBar extends SeekBar {
3      public VerticalSeekBar(Context context) {
4          super(context);
5      }
6      public VerticalSeekBar(Context context, AttributeSet attrs, int defStyle) {
7          super(context, attrs, defStyle);
8      }
9      public VerticalSeekBar(Context context, AttributeSet attrs) {
10         super(context, attrs);
11     }
12     protected void onSizeChanged(int w, int h, int oldw, int oldh) {
13         super.onSizeChanged(h, w, oldh, oldw);
14     }
15     @Override
16     protected synchronized void onMeasure(int widthMeasureSpec,
17             int heightMeasureSpec) {
18         super.onMeasure(heightMeasureSpec, widthMeasureSpec);
19         setMeasuredDimension(getMeasuredHeight(), getMeasuredWidth());
20     }
21     protected void onDraw(Canvas c) {
22         c.rotate(-90);
23         c.translate(-getHeight(), 0);
24         super.onDraw(c);
25     }
26     @Override
27     public synchronized void setProgress(int progress) {
28         super.setProgress(progress);
29         onSizeChanged(getWidth(), getHeight(), 0, 0);
30     }
31     @Override
32     public boolean onTouchEvent(MotionEvent event) {
33         if(!isEnabled()) {
34             return false;
35         }
36         switch(event.getAction()) {
37             case MotionEvent.ACTION_DOWN:
38             case MotionEvent.ACTION_MOVE:
39             case MotionEvent.ACTION_UP:
40                 setProgress((int) ((int) getMax() - (getMax() * event.getY() /
41                         getHeight())));
42                 break;
43             default:
44                 return super.onTouchEvent(event);
45         }
46         return true;
47     }
48 }
```

【任务 9-15】 画面尺寸菜单逻辑代码

【任务分析】

当横屏播放视频时,点击屏蔽右上角的"更多"图标会弹出一个选择画面尺寸的菜单,画面尺寸菜单中的选项有满屏、100%、75%以及50%,点击其中任意一项就会使播放界面缩小或放大。

【任务实施】

(1) 创建 PlayTopPopupWindow 类。在 com.itheima.topline.view 文件夹中创建一个 PlayTopPopupWindow 类,在该类中创建一个 setScreenSizeCheckLister() 方法,用于监听"选择画面尺寸"界面中按钮的点击事件。

(2) 设置画面尺寸菜单。在 PlayTopPopupWindow 类中创建一个 showAsDropDown() 方法,用于设置画面尺寸菜单的显示,具体代码如文件 9-22 所示。

【文件 9-22】 PlayTopPopupWindow.java

```java
package com.itheima.topline.view;
/**
 * 弹出菜单
 */
public class PlayTopPopupWindow {
    private PopupWindow popupWindow;
    private RadioGroup rgScreenSize;
    public PlayTopPopupWindow(Context context, int height) {
        View view=LayoutInflater.from(context).inflate(
                                    R.layout.play_top_menu, null);
        rgScreenSize=findById(R.id.rg_screensize, view);
        popupWindow=new PopupWindow(view, height * 2 / 3, height);
        popupWindow.setBackgroundDrawable(new ColorDrawable(
                                    Color.argb(178, 0, 0, 0)));
    }
    public void setScreenSizeCheckLister(RadioGroup.OnCheckedChangeListener
    listener) {
        rgScreenSize.setOnCheckedChangeListener(listener);
    }
    public void showAsDropDown(View parent) {
        popupWindow.showAtLocation(parent, Gravity.RIGHT, 0, 0);
        popupWindow.setFocusable(true);
        popupWindow.setOutsideTouchable(true);
        popupWindow.update();
    }
    public void dismiss() {
        popupWindow.dismiss();
    }
    @SuppressWarnings("unchecked")
    private<T extends View>T findById(int resId, View view) {
```

```
31          return (T) view.findViewById(resId);
32      }
33 }
```

【任务 9-16】 视频清晰度菜单逻辑代码

【任务分析】

当横屏播放视频时,点击屏幕右下角的"清晰度"按钮会弹出"显示高清和清晰"两个选项的菜单,点击"高清"或"清晰"选项界面会加载对应的视频清晰度。

【任务实施】

(1) 创建 PopMenu 类。在 com.itheima.topline.view 包中创建一个 PopMenu 类并实现 OnItemClickListener 接口,具体代码如文件 9-23 所示。

【文件 9-23】 PopMenu.java

```
1  package com.itheima.topline.view;
2  /**
3   * 弹出菜单
4   */
5  public class PopMenu implements OnItemClickListener {
6      public interface OnItemClickListener {
7          public void onItemClick(int position);
8      }
9      private ArrayList<String> itemList;
10     private Context context;
11     private PopupWindow popupWindow;
12     private ListView listView;
13     private OnItemClickListener listener;
14     private int checkedPosition;
15     public PopMenu(Context context, int resid, int checkedPosition, int height) {
16         this.context=context;
17         this.checkedPosition=checkedPosition;
18         itemList=new ArrayList<String>();
19         RelativeLayout view=new RelativeLayout(context);
20         listView=new ListView(context);
21         listView.setPadding(0, ParamsUtils.dpToPx(context, 3), 0, ParamsUtils.
22         dpToPx(context, 3));
23         view.addView(listView, new LayoutParams(LayoutParams.WRAP_CONTENT,
24                                  LayoutParams.WRAP_CONTENT));
25         listView.setAdapter(new PopAdapter());
26         listView.setOnItemClickListener(this);
27         popupWindow=new PopupWindow(view, context.getResources().
28                 getDimensionPixelSize(R.dimen.popmenu_width), height);
29         popupWindow.setBackgroundDrawable(new ColorDrawable(
30                              Color.argb(178, 0, 0, 0)));
31     }
```

```java
32      @Override
33      public void onItemClick(AdapterView<?>parent, View view, int position,
34      long id) {
35          if(listener !=null) {
36              listener.onItemClick(position);
37              checkedPosition=position;
38              listView.invalidate();
39          }
40          dismiss();
41      }
42      public void setOnItemClickListener(OnItemClickListener listener) {
43          this.listener=listener;
44      }
45      public void addItems(String[] items) {
46          for(String s : items)
47              itemList.add(s);
48      }
49      public void addItem(String item) {
50          itemList.add(item);
51      }
52      public void showAsDropDown(View parent) {
53          popupWindow.showAsDropDown(parent, parent.getWidth() / 2 * -1,
54          context.getResources().getDimensionPixelSize(R.dimen.popmenu_yoff));
55          popupWindow.setFocusable(true);
56          popupWindow.setOutsideTouchable(true);
57          popupWindow.update();
58      }
59      public void dismiss() {
60          popupWindow.dismiss();
61      }
62      private final class PopAdapter extends BaseAdapter {
63          @Override
64          public int getCount() {
65              return itemList.size();
66          }
67          @Override
68          public Object getItem(int position) {
69              return itemList.get(position);
70          }
71          @Override
72          public long getItemId(int position) {
73              return position;
74          }
75          @Override
76          public View getView(int position, View convertView, ViewGroup parent) {
77              RelativeLayout layoutView=new RelativeLayout(context);
78              TextView textView=new TextView(context);
79              textView.setTextSize(13);
80              textView.setText(itemList.get(position));
81              textView.setTag(position);
```

```
82              if(checkedPosition==position || itemList.size()==1) {
83                  textView.setTextColor(context.getResources().getColor(
84                          R.color.rb_text_check));
85              } else {
86                  textView.setTextColor(Color.WHITE);
87              }
88              LayoutParams params=new LayoutParams(LayoutParams.WRAP_CONTENT,
89              LayoutParams.WRAP_CONTENT);
90              params.addRule(RelativeLayout.CENTER_IN_PARENT);
91              layoutView.addView(textView, params);
92              layoutView.setMinimumHeight(ParamsUtils.dpToPx(context, 26));
93              return layoutView;
94          }
95      }
96  }
```

（2）修改 dimens.xml 文件。由于 PopMenu.java 文件中需要设置 PopMenu 菜单的宽度与高度，因此需要在 res/values 文件夹的 dimens.xml 文件中添加如下代码：

```
<dimen name="popmenu_width">100dp</dimen>
<dimen name="popmenu_yoff">5dp</dimen>
```

【任务 9-17】 "视频详情"界面逻辑代码

【任务分析】

"视频详情"界面主要展示视频播放、视频简介以及视频目录列表，当点击"视频简介"按钮时，将展示视频简介的布局；当点击"视频目录列表"按钮时，将展示视频目录列表布局。视频简介与视频目录列表的数据是从"视频列表"界面传递过来的，当用户点击视频目录中的任意一条时，"视频详情"界面上方都会加载播放对应的视频，同时点击正在播放的视频右下角的"全屏"按钮会横屏全屏播放视频。该界面的视频播放是调用第三方 CC 视频播放器播放的。

【任务实施】

（1）获取界面控件。在 VideoDetailActivity 中创建界面控件的初始化方法 initViewPager()，用于获取"视频简介"与"视频目录"界面所要用到的控件。

（2）创建视频播放的方法。在 VideoDetailActivity 中创建一个 VideoPlay()方法，在该方法中调用 setVideoPlayInfo()方法，并以传递的视频 Id 以及从 CC 平台上获取的 USERID 与 API_KEY 为参数实现视频播放。由于该类代码非常多，因此只给出关键代码，读者可以从源代码中复制该类，无须手动编写，理解思路即可，具体代码如文件 9-24 所示。

【文件 9-24】 VideoDetailActivity.java

```
1  public class VideoDetailActivity extends AppCompatActivity {
2      ...
3      private void initViewPager() {
```

```java
4       mViewPager=(ViewPager) findViewById(R.id.vp_view);
5         mTabLayout= (TabLayout) findViewById(R.id.tabs);
6         mInflater=LayoutInflater.from(this);
7         view1=mInflater.inflate(R.layout.video_detail_viewpager1, null);
8         view2=mInflater.inflate(R.layout.video_detail_viewpager2, null);
9         tv_intro= (TextView) view1.findViewById(R.id.tv_intro);
10        tv_intro.setText(intro);
11        lv_list=(ListView) view2.findViewById(R.id.lv_list);
12        videoDetailListAdapter=new VideoDetailListAdapter(VideoDetailActivity.this,
13        new VideoDetailListAdapter.OnSelectListener() {
14            @Override
15            public void onSelect(int position, ImageView iv) {
16                //设置适配器的选中项
17                videoDetailListAdapter.setSelectedPosition(position);
18                VideoDetailBean bean=videoList.get(position);
19                String videoId=bean.getVideo_id();
20                videoIdText.setText(bean.getVideo_name());
21                videoDetailListAdapter.notifyDataSetChanged();    //更新列表框
22                if(TextUtils.isEmpty(videoId)) {
23                    Toast.makeText(VideoDetailActivity.this,
24                    "本地没有此视频,暂时无法播放", Toast.LENGTH_SHORT).show();
25                    return;
26                } else {
27                    VideoPlay(videoId);
28                }
29            }
30        });
31        //没有点击视频时默认播放第一个视频
32        videoDetailListAdapter.setSelectedPosition(0);
33        videoDetailListAdapter.setData(videoList);
34        lv_list.setAdapter(videoDetailListAdapter);
35        //添加页卡视图
36        mViewList.add(view1);
37        mViewList.add(view2);
38        //添加页卡标题
39        mTitleList.add("视频简介");
40        mTitleList.add("视频目录");
41        mTabLayout.setTabMode(TabLayout.MODE_FIXED);    //设置 Tab 模式,当前为默认模式
42        //添加 Tab 选项卡
43        mTabLayout.addTab(mTabLayout.newTab().setText(mTitleList.get(0)));
44        mTabLayout.addTab(mTabLayout.newTab().setText(mTitleList.get(1)));
45        VideoDetailPagerAdapter mAdapter=new VideoDetailPagerAdapter(
46                                            mViewList, mTitleList);
47        mViewPager.setAdapter(mAdapter);           //给 ViewPager 设置适配器
48        mTabLayout.setupWithViewPager(mViewPager); //给 TabLayout 设置关联
49        mTabLayout.setTabsFromPagerAdapter(mAdapter); //给 Tabs 设置适配器
50    }
51    private void VideoPlay(String videoId) {
52        isPrepared=false;
53        player.pause();
```

```
54          player.stop();
55          bufferProgressBar.setVisibility(View.VISIBLE);
56          player.reset();
57          player.setDefaultDefinition(defaultDefinition);
58          player.setVideoPlayInfo(videoId, USERID, API_KEY, VideoDetailActivity.this);
59          player.setDisplay(surfaceHolder);
60          application.getDRMServer().reset();
61          player.prepareAsync();
62       }
63       ...
64    }
```

(3) 修改"视频列表"界面 Adapter。由于点击"视频列表"界面的条目会跳转到"视频详情"界面,因此需要在"视频列表"界面的 Adapter 中添加跳转代码。首先要找到文件 9-5 中的 onBindViewHolder()方法,在该方法的 onClick()方法中添加如下代码:

```
Intent intent=new Intent(context, VideoDetailActivity.class);
intent.putExtra("intro", bean.getIntro());
intent.putExtra("videoDetailList", (Serializable) bean.getVideoDetailList());
context.startActivity(intent);
```

(4) 修改 dimens.xml 文件。由于在"视频详情"界面中需要设置菜单的高度,因此需要在 res/values 文件夹的 dimens.xml 文件中添加如下代码:

```
<dimen name="popmenu_height">60dp</dimen>
```

(5) 修改清单文件。"视频详情"界面在播放视频时会根据物理方向的传感器确定界面是横屏还是竖屏,因此需要在清单文件中设置 VideoDetailActivity 的属性 screenOrientation 的值为 sensor,为了避免切换屏幕方向时重新调用各个生命周期,需要设置属性 configChanges 的值为 screenSize|keyboardHidden|orientation,具体代码如下所示:

```
<activity
    android:name=".activity.VideoDetailActivity"
    android:configChanges="screenSize|keyboardHidden|orientation"
    android:screenOrientation="sensor" />
```

由于"视频详情"界面在播放视频时需要判断当前的网络状态,因此需要在清单文件中添加允许查看当前网络状态的权限,具体代码如下所示:

```
<!--CC 视频播放需要的权限-->
<uses-permission android:name="android.permission.ACCESS_NETWORK_STATE" />
```

9.3 本章小结

本章主要讲解了视频模块,包括"视频列表"与"视频详情"。读者需要明确各个模块之间的关系,对各界面之间的跳转逻辑有清晰的认识。同时需要掌握 CC 视频播放器的使用,视频播放的逻辑代码较为复杂,需要读者认真地分析与研究。

【思考题】

1. 如何实现"视频列表"界面?
2. 如何实现视频播放功能?

第 10 章

"我"模块（一）

学习目标

- 掌握 SQLite 数据库的使用，能够使用数据库存放用户信息
- 掌握"我"界面开发，能够展示用户基本信息以及该界面的功能
- 掌握"登录""注册"界面的开发，实现用户登录注册功能
- 掌握"个人资料"以及"修改"界面的开发，实现用户信息的展示与修改功能

"我"模块（一）主要以用户注册、登录以及管理用户信息为主。当用户注册账号并登录成功后，可以修改用户的个人信息，如头像、性别、昵称、签名等，本章将针对"我"模块（一）进行详细讲解。

10.1 创建数据库

任务综述

思政材料 10

根据"我"界面设计图可知，"我"模块中包含了用户信息，为了便于后续对用户信息进行增、删、改、查的操作，需要创建一个数据库把这些信息存储起来。

【任务 10-1】 创建 SQLite 数据库

【任务分析】

由于"我"模块涉及用户信息，同时后续会对用户信息进行操作，因此需要创建一个数据库与个人信息表，把用户信息保存到数据库中。

【任务实施】

（1）创建 SQLiteHelper 类。选中 com.itheima.topline 包，在该包下创建一个 sqlite 包。在 sqlite 包中创建一个 Java 类，命名为 SQLiteHelper 并继承 SQLiteOpenHelper 类，同时重写 onCreate()方法，该类用于创建 topline.db 数据库。

（2）创建用户信息表。由于个人资料界面的数据需要单独的一个表储存，因此需要在 onCreate()方法中通过执行一条建表的 SQL 语句创建用户信息表，具体代码如文件 10-1 所示。

【文件 10-1】 SQLiteHelper.java

```
1   package com.itheima.topline.sqlite;
2   public class SQLiteHelper extends SQLiteOpenHelper {
3       private static final int DB_VERSION=1;
4       public static String DB_NAME="topline.db";
5       public static final String U_USERINFO="userinfo";        //用户信息
6       public SQLiteHelper(Context context) {
7           super(context, DB_NAME, null, DB_VERSION);
8       }
9       @Override
10      public void onCreate(SQLiteDatabase db) {
11          /**
12           * 创建用户信息表
13           */
14          db.execSQL("CREATE TABLE IF NOT EXISTS "+U_USERINFO+"( "
15                  +"_id INTEGER PRIMARY KEY AUTOINCREMENT, "
16                  +"userName VARCHAR, "                //用户名
17                  +"nickName VARCHAR, "                //昵称
18                  +"sex VARCHAR, "                     //性别
19                  +"signature VARCHAR,"       //签名
20                  +"head VARCHAR "            //头像
21                  +")");
22      }
23      /**
24       * 当数据库版本号增加时才会调用此方法
25       */
26      @Override
27      public void onUpgrade(SQLiteDatabase db, int oldVersion, int newVersion) {
28          db.execSQL("DROP TABLE IF EXISTS "+U_USERINFO);
29          onCreate(db);
30      }
31  }
```

【任务 10-2】 创建 DBUtils 类

【任务分析】

当读取用户资料或者对用户信息进行更改时，需要对数据库进行操作，因此需要创建一个 DBUtils 工具类专门用于操作数据库。

【任务实施】

在 com.itheima.topline.utils 包中创建一个 Java 类，命名为 DBUtils，具体代码如文件 10-2 所示。

【文件 10-2】 DBUtils.java

```
1   package com.itheima.topline.utils;
2   public class DBUtils {
3       private static DBUtils instance=null;
```

```
4    private static SQLiteHelper helper;
5    private static SQLiteDatabase db;
6    public DBUtils(Context context) {
7        helper=new SQLiteHelper(context);
8        db=helper.getWritableDatabase();
9    }
10   public static DBUtils getInstance(Context context) {
11       if(instance==null) {
12           instance=new DBUtils(context);
13       }
14       return instance;
15   }
16 }
```

【任务 10-3】 创建 UserBean

【任务分析】

"我"模块涉及用户信息,用户具有用户名、昵称、性别等属性,为了便于后续对这些属性进行操作,需要创建一个 UserBean 类存放这些属性。

【任务实施】

在 com.itheima.topline.bean 包中创建一个 Java 类,命名为 UserBean。在该类中创建用户所需的属性,具体代码如文件 10-3 所示。

【文件 10-3】 UserBean.java

```
1  package com.itheima.topline.bean;
2  public class UserBean {
3      private String userName;       //用户名
4      private String nickName;       //昵称
5      private String sex;            //性别
6      private String signature;      //签名
7      private String head;           //头像
8      public String getUserName() {
9          return userName;
10     }
11     public void setUserName(String userName) {
12         this.userName=userName;
13     }
14     public String getNickName() {
15         return nickName;
16     }
17     public void setNickName(String nickName) {
18         this.nickName=nickName;
19     }
20     public String getSex() {
21         return sex;
22     }
```

```
23     public void setSex(String sex) {
24         this.sex=sex;
25     }
26     public String getSignature() {
27         return signature;
28     }
29     public void setSignature(String signature) {
30         this.signature=signature;
31     }
32     public String getHead() {
33         return head;
34     }
35     public void setHead(String head) {
36         this.head=head;
37     }
38 }
```

10.2 "我"

任务综述

根据"我"界面设计图可知,该界面包含了用户头像、用户名、日历、星座、涂鸦、地图、收藏以及设置。当用户处于登录状态时,点击用户头像会跳转到"个人资料"界面;当用户处于未登录状态时,点击用户头像会跳转到"登录"界面。

【任务 10-4】 "我"界面

【任务分析】

"我"界面需要显示头像、用户名、日历、星座、涂鸦、地图、收藏条目以及设置条目,界面效果如图 10-1 所示。

图 10-1 "我"界面

【任务实施】

(1) 创建"我"界面。在 res/layout 文件夹中创建一个布局文件 fragment_me.xml。

(2) 导入界面图片。将"我"界面所需的图片 my_bg.png、calendar_icon.png、constellation_icon.png、scraw_icon.png、map_icon.png、collection_icon.png、iv_right_arrow.png、setting_icon.png、default_head.png 导入 drawable-hdpi 文件夹。

(3) 添加 circleimageview 库。由于"我"界面中的头像用到了 circleimageview 库中的 CircleImageView 控件,因此需要添加 circleimageview 库到该项目中,在

Android Studio 中,选中项目,右击选择 Open Module Settings 选项后选择 Dependencies 选项卡,单击右上角的绿色加号,选择 Library dependency 选项,把 de. hdodenhof:circleimageview:1.3.0 库加入主项目中。

注意：如果在 Library dependency 选项中找不到该库,则可以直接在该项目的 build.gradle 文件中添加如下代码。

```
compile 'de.hdodenhof:circleimageview:1.3.0'
```

（4）添加 cardview-v7 库。由于"我"界面中需要显示类似卡片的样式,因此需要在 Android Studio 中,选中项目,右击选择 Open Module Settings 选项后选择 Dependencies 选项卡,单击右上角的绿色加号并选择 Library dependency 选项,然后找到 com. android. support:cardview-v7 库并添加到项目中。

（5）放置界面控件。在布局文件中,放置一个 ImageView 控件用于显示红色背景图片；一个 Toolbar 控件用于显示上下滑动的标题栏；一个 CircleImageView 控件用于显示头像；在该布局文件中,通过＜include＞标签将 layout_content_me. xml(局部布局)引入,具体代码如文件 10-4 所示。

【文件 10-4】 fragment_me. xml

```
1    <?xml version="1.0" encoding="utf-8"?>
2    <android.support.design.widget.CoordinatorLayout
3        xmlns:android="http://schemas.android.com/apk/res/android"
4        xmlns:app="http://schemas.android.com/apk/res-auto"
5        android:layout_width="match_parent"
6        android:layout_height="match_parent"
7        android:background="@android:color/white">
8        <android.support.design.widget.AppBarLayout
9            android:id="@+id/appbar"
10           android:layout_width="match_parent"
11           android:layout_height="wrap_content"
12           android:fitsSystemWindows="true"
13           android:theme="@style/ThemeOverlay.AppCompat.Dark.ActionBar">
14           <android.support.design.widget.CollapsingToolbarLayout
15               android:id="@+id/collapsing_tool_bar"
16               android:layout_width="match_parent"
17               android:layout_height="match_parent"
18               android:fitsSystemWindows="true"
19               app:contentScrim="?attr/colorPrimary"
20               app:expandedTitleGravity="bottom|center_horizontal"
21               app:layout_scrollFlags="scroll|exitUntilCollapsed|snap">
22               <ImageView
23                   android:layout_width="match_parent"
24                   android:layout_height="220dp"
25                   android:scaleType="centerCrop"
26                   android:src="@drawable/my_bg"
27                   app:layout_collapseMode="parallax"
28                   app:layout_collapseParallaxMultiplier="0.6" />
```

```
29          <android.support.v7.widget.Toolbar
30              android:id="@+id/tool_bar"
31              android:layout_width="match_parent"
32              android:layout_height="?attr/actionBarSize"
33              app:layout_collapseMode="pin"
34              app:navigationIcon="@android:color/transparent"
35              app:theme="@style/ThemeOverlay.AppCompat.Dark" />
36      </android.support.design.widget.CollapsingToolbarLayout>
37  </android.support.design.widget.AppBarLayout>
38  <android.support.v4.widget.NestedScrollView
39      android:layout_width="match_parent"
40      android:layout_height="match_parent"
41      app:layout_behavior="@string/appbar_scrolling_view_behavior">
42      <include layout="@layout/layout_content_me" />
43  </android.support.v4.widget.NestedScrollView>
44  <!--layout_anchor 属性 5.0 以上需要设置为 CollapsingToolbarLayout 不然头像会被覆盖-->
45  <de.hdodenhof.circleimageview.CircleImageView
46      android:id="@+id/iv_avatar"
47      android:layout_width="70dp"
48      android:layout_height="70dp"
49      app:border_color="@android:color/white"
50      app:border_width="1dp"
51      app:layout_anchor="@id/collapsing_tool_bar"
52      app:layout_anchorGravity="center"
53      app:layout_behavior="com.itheima.topline.view.AvatarBehavior" />
54  </android.support.design.widget.CoordinatorLayout>
```

（6）放置 layout_content_me.xml 文件中的控件。在 res/layout 文件夹中创建一个 layout_content_me.xml 文件。在该布局文件中，放置 8 个 ImageView 控件，其中 4 个分别用于显示日历图标、星座图标、涂鸦图标以及地图图标，2 个用于显示收藏图标与设置图标，2 个用于显示收藏与设置条目后面的箭头；放置 6 个 TextView 控件，其中 4 个分别用于显示日历文本、星座文本、涂鸦文本以及地图文本；2 个分别用于显示收藏与设置的文本；放置 4 个 View 控件，其中 3 个用于分隔日历、星座、涂鸦、地图，1 个用于分隔收藏与设置条目，具体代码如文件 10-5 所示。

【文件 10-5】 layout_content_me.xml

```
1   <?xml version="1.0" encoding="utf-8"?>
2   <LinearLayout xmlns:android="http://schemas.android.com/apk/res/android"
3       xmlns:app="http://schemas.android.com/apk/res-auto"
4       android:layout_width="match_parent"
5       android:layout_height="wrap_content"
6       android:orientation="vertical">
7       <android.support.v7.widget.CardView
8           android:layout_width="match_parent"
9           android:layout_height="wrap_content"
10          android:layout_margin="8dp"
11          app:cardCornerRadius="5dp"
12          app:cardElevation="3dp"
13          app:cardPreventCornerOverlap="false"
```

```xml
14        app:cardUseCompatPadding="true">
15        <LinearLayout
16            android:layout_width="match_parent"
17            android:layout_height="wrap_content"
18            android:orientation="horizontal"
19            android:paddingBottom="8dp"
20            android:paddingTop="8dp">
21            <LinearLayout
22                android:id="@+id/ll_calendar"
23                android:layout_width="0dp"
24                android:layout_height="wrap_content"
25                android:layout_weight="1"
26                android:gravity="center_horizontal"
27                android:orientation="vertical">
28                <ImageView
29                    android:layout_width="@dimen/star_img_height"
30                    android:layout_height="@dimen/star_img_height"
31                    android:scaleType="fitXY"
32                    android:src="@drawable/calendar_icon" />
33                <TextView
34                    android:layout_width="wrap_content"
35                    android:layout_height="wrap_content"
36                    android:layout_marginTop="4dp"
37                    android:text="日历"
38                    android:textColor="@color/starTextColor"
39                    android:textSize="@dimen/star_text_size" />
40            </LinearLayout>
41            <View
42                android:layout_width="1dp"
43                android:layout_height="fill_parent"
44                android:background="@color/divider_line_color" />
45            <LinearLayout
46                android:id="@+id/ll_constellation"
47                android:layout_width="0dp"
48                android:layout_height="wrap_content"
49                android:layout_weight="1"
50                android:gravity="center_horizontal"
51                android:orientation="vertical">
52                <ImageView
53                    android:layout_width="@dimen/star_img_height"
54                    android:layout_height="@dimen/star_img_height"
55                    android:scaleType="fitXY"
56                    android:src="@drawable/constellation_icon" />
57                <TextView
58                    android:layout_width="wrap_content"
59                    android:layout_height="wrap_content"
60                    android:layout_marginTop="4dp"
61                    android:text="星座"
62                    android:textColor="@color/starTextColor"
63                    android:textSize="@dimen/star_text_size" />
```

```xml
            </LinearLayout>
            <View
                android:layout_width="1dp"
                android:layout_height="fill_parent"
                android:background="@color/divider_line_color" />
            <LinearLayout
                android:id="@+id/ll_scraw"
                android:layout_width="0dp"
                android:layout_height="wrap_content"
                android:layout_weight="1"
                android:gravity="center_horizontal"
                android:orientation="vertical">
                <ImageView
                    android:layout_width="@dimen/star_img_height"
                    android:layout_height="@dimen/star_img_height"
                    android:scaleType="fitXY"
                    android:src="@drawable/scraw_icon" />
                <TextView
                    android:layout_width="wrap_content"
                    android:layout_height="wrap_content"
                    android:layout_marginTop="4dp"
                    android:text="涂鸦"
                    android:textColor="@color/starTextColor"
                    android:textSize="@dimen/star_text_size" />
            </LinearLayout>
            <View
                android:layout_width="1dp"
                android:layout_height="fill_parent"
                android:background="@color/divider_line_color" />
            <LinearLayout
                android:id="@+id/ll_map"
                android:layout_width="0dp"
                android:layout_height="wrap_content"
                android:layout_weight="1"
                android:gravity="center_horizontal"
                android:orientation="vertical">
                <ImageView
                    android:layout_width="@dimen/star_img_height"
                    android:layout_height="@dimen/star_img_height"
                    android:scaleType="fitXY"
                    android:src="@drawable/map_icon" />
                <TextView
                    android:layout_width="wrap_content"
                    android:layout_height="wrap_content"
                    android:layout_marginTop="4dp"
                    android:text="地图"
                    android:textColor="@color/starTextColor"
                    android:textSize="@dimen/star_text_size" />
            </LinearLayout>
        </LinearLayout>
```

```xml
114    </android.support.v7.widget.CardView>
115    <android.support.v7.widget.CardView
116        android:layout_width="match_parent"
117        android:layout_height="wrap_content"
118        android:layout_marginLeft="8dp"
119        android:layout_marginRight="8dp"
120        app:cardCornerRadius="5dp"
121        app:cardElevation="3dp"
122        app:cardPreventCornerOverlap="false"
123        app:cardUseCompatPadding="true">
124        <LinearLayout
125            android:layout_width="fill_parent"
126            android:layout_height="wrap_content"
127            android:layout_marginLeft="8dp"
128            android:layout_marginRight="8dp"
129            android:orientation="vertical">
130            <RelativeLayout
131                android:id="@+id/rl_collection"
132                android:layout_width="fill_parent"
133                android:layout_height="50dp"
134                android:layout_marginLeft="8dp"
135                android:layout_marginRight="8dp">
136                <ImageView
137                    android:id="@+id/iv_collec"
138                    android:layout_width="20dp"
139                    android:layout_height="20dp"
140                    android:layout_centerVertical="true"
141                    android:scaleType="fitXY"
142                    android:src="@drawable/collection_icon" />
143                <TextView
144                    android:layout_width="wrap_content"
145                    android:layout_height="wrap_content"
146                    android:layout_centerVertical="true"
147                    android:layout_marginLeft="8dp"
148                    android:layout_toRightOf="@id/iv_collec"
149                    android:text="收藏"
150                    android:textColor="@color/starTextColor"
151                    android:textSize="@dimen/star_text_size" />
152                <ImageView
153                    android:layout_width="10dp"
154                    android:layout_height="10dp"
155                    android:layout_alignParentRight="true"
156                    android:layout_centerVertical="true"
157                    android:src="@drawable/iv_right_arrow" />
158            </RelativeLayout>
159            <View
160                android:layout_width="fill_parent"
161                android:layout_height="1dp"
162                android:background="@color/divider_line_color" />
163            <RelativeLayout
```

```
164                     android:id="@+id/rl_setting"
165                     android:layout_width="fill_parent"
166                     android:layout_height="50dp"
167                     android:layout_marginLeft="8dp"
168                     android:layout_marginRight="8dp">
169                     <ImageView
170                         android:id="@+id/iv_setting"
171                         android:layout_width="20dp"
172                         android:layout_height="20dp"
173                         android:layout_centerVertical="true"
174                         android:src="@drawable/setting_icon" />
175                     <TextView
176                         android:layout_width="wrap_content"
177                         android:layout_height="wrap_content"
178                         android:layout_centerVertical="true"
179                         android:layout_marginLeft="8dp"
180                         android:layout_toRightOf="@id/iv_setting"
181                         android:text="设置"
182                         android:textColor="@color/starTextColor"
183                         android:textSize="@dimen/star_text_size" />
184                     <ImageView
185                         android:layout_width="10dp"
186                         android:layout_height="10dp"
187                         android:layout_alignParentRight="true"
188                         android:layout_centerVertical="true"
189                         android:src="@drawable/iv_right_arrow" />
190                 </RelativeLayout>
191             </LinearLayout>
192     </android.support.v7.widget.CardView>
193 </LinearLayout>
```

(7) 修改 colors.xml 文件与 dimens.xml 文件。根据以上布局文件,可知需要在 res/values 文件夹的 colors.xml 文件中添加如下代码:

```
<color name="starTextColor">#999999</color>
<color name="divider_line_color">#ebebeb</color>
<color name="colorPrimary">#eb413d</color>
```

需要注意的是,在 colors.xml 文件中默认会有 colorPrimary 颜色,仅修改其值即可。
在 res/values 文件夹的 dimens.xml 文件中添加如下代码:

```
<dimen name="star_img_height">25dp</dimen>
<dimen name="star_text_size">14sp</dimen>
<dimen name="image_width">120dp</dimen>
<dimen name="spacing_normal">16dp</dimen>
<dimen name="avatar_final_size">40dp</dimen>
<dimen name="avatar_final_x">8dp</dimen>
<dimen name="toolbar_height">56dp</dimen>
```

（8）添加一个自定义 AvatarBehavior。由于向上推动"我"界面时，头像会向上移动到屏幕左上角，向下拉动"我"界面时，头像会移动到它原的位置，因此需要自定义一个 AvatarBehavior 来设置头像移动的轨迹以及移动的最终位置，在 com.itheima.topline.view 包中创建一个 AvatarBehavior 类并继承 CoordinatorLayout.Behavior<CircleImageView> 类，具体代码如文件 10-6 所示。

【文件 10-6】 AvatarBehavior.java

```java
1   package com.itheima.topline.view;
2   import de.hdodenhof.circleimageview.CircleImageView;
3   /**
4    * 头像 Behavior
5    */
6   public class AvatarBehavior extends CoordinatorLayout.Behavior<CircleImageView>
7   {
8       private static final float ANIM_CHANGE_POINT=0.2f;       //缩放动画变化的支点
9       private Context mContext;
10      private int mTotalScrollRange;          //整个滚动的范围
11      private int mAppBarHeight;              //AppBarLayout 高度
12      private int mAppBarWidth;               //AppBarLayout 宽度
13      private int mOriginalSize;              //控件原始大小
14      private int mFinalSize;                 //控件最终大小
15      private float mScaleSize;               //控件最终缩放的尺寸,设置坐标值需要算上该值
16      private float mOriginalX;               //原始 X 坐标
17      private float mFinalX;                  //最终 X 坐标
18      private float mOriginalY;               //原始 Y 坐标
19      private float mFinalY;                  //最终 Y 坐标
20      private int mToolBarHeight;             //ToolBar 高度
21      private float mAppBarStartY;            //AppBar 的起始 Y 坐标
22      private float mPercent;                 //滚动执行百分比[0~1]
23      private DecelerateInterpolator mMoveYInterpolator;    //Y轴移动插值器
24      private AccelerateInterpolator mMoveXInterpolator;    //X轴移动插值器
25      //最终变换的视图,因为在 5.0 以上 AppBarLayout 在收缩到最终状态时会覆盖变换后的视图,
26      //所以添加一个最终显示的图片
27      private CircleImageView mFinalView;
28      private int mFinalViewMarginBottom;    //最终变换的视图距离底部的大小
29      public AvatarBehavior(Context context, AttributeSet attrs) {
30          super(context, attrs);
31          mContext=context;
32          mMoveYInterpolator=new DecelerateInterpolator();
33          mMoveXInterpolator=new AccelerateInterpolator();
34          if(attrs !=null) {
35              TypedArray a=mContext.obtainStyledAttributes(attrs,
36                      R.styleable.AvatarImageBehavior);
37              mFinalSize=(int) a.getDimension(R.styleable.
38                      AvatarImageBehavior_finalSize, 0);
39              mFinalX=a.getDimension(R.styleable.AvatarImageBehavior_finalX, 0);
40              mToolBarHeight= (int) a.getDimension(R.styleable.
41                      AvatarImageBehavior_toolBarHeight, 0);
42              a.recycle();
43          }
44      }
```

```java
45      @Override
46      public boolean layoutDependsOn(CoordinatorLayout parent, CircleImageView
47      child, View dependency) {
48          return dependency instanceof AppBarLayout;
49      }
50      @Override
51      public boolean onDependentViewChanged(CoordinatorLayout parent, CircleImageView
52      child, View dependency) {
53          if(dependency instanceof AppBarLayout) {
54              _initVariables(child, dependency);
55              mPercent=(mAppBarStartY -dependency.getY()) * 1.0f / mTotalScrollRange;
56              float percentY=mMoveYInterpolator.getInterpolation(mPercent);
57              AnimHelper.setViewY(child, mOriginalY, mFinalY -mScaleSize, percentY);
58              if(mPercent>ANIM_CHANGE_POINT) {
59                  float scalePercent=(mPercent -ANIM_CHANGE_POINT) / (1 -
60                  ANIM_CHANGE_POINT);
61                  float percentX=mMoveXInterpolator.getInterpolation(scalePercent);
62                  AnimHelper.scaleView(child, mOriginalSize, mFinalSize, scalePercent);
63                  AnimHelper.setViewX(child, mOriginalX, mFinalX -mScaleSize, percentX);
64              } else {
65                  AnimHelper.scaleView(child, mOriginalSize, mFinalSize, 0);
66                  AnimHelper.setViewX(child, mOriginalX, mFinalX -mScaleSize, 0);
67              }
68              if(mFinalView !=null) {
69                  if(percentY==1.0f) {
70                      //滚动到顶时才显示
71                      mFinalView.setVisibility(View.VISIBLE);
72                  } else {
73                      mFinalView.setVisibility(View.GONE);
74                  }
75              }
76          } else if(Build.VERSION.SDK_INT>=Build.VERSION_CODES.LOLLIPOP &&
77                  dependency instanceof CollapsingToolbarLayout) {
78              //大于5.0才生成新的、最终的头像,因为5.0以上的AppBarLayout会覆盖变换后的头像
79              if(mFinalView==null && mFinalSize !=0 && mFinalX !=0 &&
80              mFinalViewMarginBottom !=0) {
81                  mFinalView=new CircleImageView(mContext);
82                  mFinalView.setVisibility(View.GONE);
83                  //添加为CollapsingToolbarLayout子视图
84                  ((CollapsingToolbarLayout) dependency).addView(mFinalView);
85                  FrameLayout.LayoutParams params= (FrameLayout.LayoutParams)
86                                                  mFinalView.getLayoutParams();
87                  params.width=mFinalSize;        //设置大小
88                  params.height=mFinalSize;
89                  params.gravity=Gravity.BOTTOM;   //设置位置,最后显示时相当于变换后的头像位置
90                  params.leftMargin= (int) mFinalX;
91                  params.bottomMargin=mFinalViewMarginBottom;
92                  mFinalView.setLayoutParams(params);
93                  mFinalView.setImageDrawable(child.getDrawable());
94                  mFinalView.setBorderColor(child.getBorderColor());
```

```java
 95             int borderWidth=(int) ((mFinalSize * 1.0f / mOriginalSize) *
 96                 child.getBorderWidth());
 97             mFinalView.setBorderWidth(borderWidth);
 98         } else {
 99             mFinalView.setImageDrawable(child.getDrawable());
100             mFinalView.setBorderColor(child.getBorderColor());
101             int borderWidth=(int) ((mFinalSize * 1.0f / mOriginalSize) *
102                 child.getBorderWidth());
103             mFinalView.setBorderWidth(borderWidth);
104         }
105     }
106     return true;
107 }
108 /**
109  * 初始化变量
110  */
111 private void _initVariables(CircleImageView child, View dependency) {
112     if(mAppBarHeight==0) {
113         mAppBarHeight=dependency.getHeight();
114         mAppBarStartY=dependency.getY();
115     }
116     if(mTotalScrollRange==0) {
117         mTotalScrollRange= ((AppBarLayout) dependency).getTotalScrollRange();
118     }
119     if(mOriginalSize==0) {
120         mOriginalSize=child.getWidth();
121     }
122     if(mFinalSize==0) {
123         mFinalSize=mContext.getResources().getDimensionPixelSize(R.dimen.
124             avatar_final_size);
125     }
126     if(mAppBarWidth==0) {
127         mAppBarWidth=dependency.getWidth();
128     }
129     if(mOriginalX==0) {
130         mOriginalX=child.getX();
131     }
132     if(mFinalX==0) {
133         mFinalX=mContext.getResources().getDimensionPixelSize(R.dimen.
134             avatar_final_x);
135     }
136     if(mOriginalY==0) {
137         mOriginalY=child.getY();
138     }
139     if(mFinalY==0) {
140         if(mToolBarHeight==0) {
141             mToolBarHeight=mContext.getResources().getDimensionPixelSize(
142                 R.dimen.toolbar_height);
143         }
144         mFinalY= (mToolBarHeight -mFinalSize) / 2+mAppBarStartY;
145     }
```

```
146            if(mScaleSize==0) {
147                mScaleSize=(mOriginalSize -mFinalSize) * 1.0f / 2;
148            }
149            if(mFinalViewMarginBottom==0) {
150                mFinalViewMarginBottom=(mToolBarHeight -mFinalSize)/2;
151            }
152        }
153    }
```

（9）添加一个 attrs.xml 文件。由于需要设置自定义控件 AvatarBehavior 中的属性值的类型，因此需要在 res/values 文件夹中创建一个 attrs.xml 文件，该文件中的 name 表示属性的名称，format 表示属性的类型，具体代码如文件 10-7 所示。

【文件 10-7】 attrs.xml

```
1  <?xml version="1.0" encoding="utf-8"?>
2  <resources>
3      <declare-styleable name="AvatarImageBehavior">
4          <attr name="finalSize" format="dimension" />
5          <attr name="finalX" format="dimension" />
6          <attr name="toolBarHeight" format="dimension" />
7          <attr name="finalYPosition" format="dimension" />
8          <attr name="startXPosition" format="dimension" />
9          <attr name="startToolbarPosition" format="dimension" />
10         <attr name="startHeight" format="dimension" />
11         <attr name="finalHeight" format="dimension" />
12     </declare-styleable>
13 </resources>
```

（10）创建 AnimHelper 类。由于需要设置 AvatarBehavior 控件的 X 轴与 Y 轴的坐标，因此在 com.itheima.topline.utils 包中创建一个 AnimHelper 类，在该类中创建 setViewX() 与 setViewY() 方法，分别用于设置 X 轴与 Y 轴坐标，具体代码如文件 10-8 所示。

【文件 10-8】 AnimHelper.java

```
1  package com.itheima.topline.utils;
2  public class AnimHelper {
3      private AnimHelper() {
4          throw new RuntimeException("AnimHelper cannot be initialized!");
5      }
6      public static void setViewX(View view, float originalX, float finalX,
7              float percent) {
8          float calcX=(finalX -originalX) * percent+originalX;
9          view.setX(calcX);
10     }
11     public static void setViewY(View view, float originalY, float finalY,
12             float percent) {
13         float calcY=(finalY -originalY) * percent+originalY;
14         view.setY(calcY);
15     }
```

```
16    public static void scaleView(View view, float originalSize, float finalSize,
17    float percent) {
18        float calcSize= (finalSize -originalSize) * percent+originalSize;
19        float caleScale=calcSize / originalSize;
20        view.setScaleX(caleScale);
21        view.setScaleY(caleScale);
22    }
23 }
```

【任务 10-5】 广播接收者

【任务分析】

由于在个人头像修改后,需要在"我"界面更新头像,因此需要创建一个广播快速更新头像。

【任务实施】

选中 com.itheima.topline 包,在该包下创建一个 receiver 包。在 receiver 包中创建一个 Java 类,命名为 UpdateUserInfoReceiver 并继承 BroadcastReceiver 类,同时重写 onReceive() 方法。具体代码如文件 10-9 所示。

【文件 10-9】 UpdateUserInfoReceiver.java

```
1   package com.itheima.topline.receiver;
2   public class UpdateUserInfoReceiver extends BroadcastReceiver {
3       public interface ACTION {
4           String UPDATE_USERINFO="update_userinfo";
5       }
6       //广播 intent 类型
7       public interface INTENT_TYPE {
8           String TYPE_NAME="intent_name";
9           String UPDATE_HEAD="update_head";        //更新头像
10      }
11      private BaseOnReceiveMsgListener onReceiveMsgListener;
12      public UpdateUserInfoReceiver(BaseOnReceiveMsgListener onReceiveMsgListener) {
13          this.onReceiveMsgListener=onReceiveMsgListener;
14      }
15      @Override
16      public void onReceive(Context context, Intent intent) {
17          onReceiveMsgListener.onReceiveMsg(context, intent);
18      }
19      public interface BaseOnReceiveMsgListener {
20          void onReceiveMsg(Context context, Intent intent);
21      }
22 }
```

【任务 10-6】 "我"界面逻辑代码

【任务分析】

在"我"界面中需要判断用户是否登录。若用户未登录,则界面上显示"点击登录",并且点击头像会跳转到"登录"界面;若用户已经登录,则界面上显示用户名,并且点击头像会跳转到"个人资料"界面。点击"收藏"条目时跳转到"收藏"界面,点击"设置"条目时跳转到"设置"界面。

【任务实施】

(1) 创建 MeFragment 类。在 com.itheima.topline.fragment 包中创建一个 Java 类,命名为 MeFragment 并实现 OnClickListener 接口。

(2) 获取界面控件。创建界面控件的初始化方法 initView(),用于获取"我"界面上所要用到的控件,并通过 readLoginStatus() 方法判断当前是否为登录状态,如果是,则需要设置对应控件的状态。同时在此方法中还需要处理控件的点击事件。

(3) 接收广播。由于个人资料界面在修改头像成功后需要及时更新"我"界面的头像,因此需要在 MeFragment 类中创建一个 receiver() 方法接收传递过来的头像信息并更新界面头像。

(4) 设置"我"界面的登录成功状态。在 MeFragment 类中创建一个 setLoginParams() 方法,用于设置登录成功后"我"界面中的头像与用户名的显示状态。

(5) 回传数据。在 MeFragment 类中重写 onActivityResult() 方法,在该方法中接收从"登录"界面(暂未创建)或者"设置"界面(暂未创建)回传过来的登录状态,从而设置"我"界面,具体代码如文件 10-10 所示。

【文件 10-10】 MeFragment.java

```
1   package com.itheima.topline.fragment;
2   public class MeFragment extends Fragment implements View.OnClickListener {
3       private LinearLayout ll_calendar, ll_constellation, ll_scraw, ll_map;
4       private RelativeLayout rl_collection, rl_setting;
5       private CircleImageView iv_avatar;
6       private View view;
7       private UpdateUserInfoReceiver updateUserInfoReceiver;
8       private IntentFilter filter;
9       private boolean isLogin=false;
10      private CollapsingToolbarLayout collapsingToolbarLayout;
11      public MeFragment() {
12      }
13      @Override
14      public View onCreateView(LayoutInflater inflater, ViewGroup container,
15                               Bundle savedInstanceState) {
16          view=inflater.inflate(R.layout.fragment_me, container, false);
17          initView(view);
18          return view;
19      }
```

```java
20    private void initView(View view) {
21        ll_calendar=(LinearLayout) view.findViewById(R.id.ll_calendar);
22        ll_constellation=(LinearLayout) view.findViewById(R.id.ll_constellation);
23        ll_scraw=(LinearLayout) view.findViewById(R.id.ll_scraw);
24        ll_map=(LinearLayout) view.findViewById(R.id.ll_map);
25        rl_collection=(RelativeLayout) view.findViewById(R.id.rl_collection);
26        rl_setting=(RelativeLayout) view.findViewById(R.id.rl_setting);
27        iv_avatar=(CircleImageView) view.findViewById(R.id.iv_avatar);
28        collapsingToolbarLayout=(CollapsingToolbarLayout) view.findViewById(
29                                              R.id.collapsing_tool_bar);
30        collapsingToolbarLayout.setExpandedTitleTextAppearance(
31                                              R.style.ToolbarTitle);
32        isLogin=UtilsHelper.readLoginStatus(getActivity());
33        setLoginParams(isLogin);
34        setListener();
35        receiver();
36    }
37    private void receiver() {
38        updateUserInfoReceiver=new UpdateUserInfoReceiver(
39                new UpdateUserInfoReceiver.BaseOnReceiveMsgListener() {
40                    @Override
41                    public void onReceiveMsg(Context context, Intent intent) {
42                        String action=intent.getAction();
43                        if(UpdateUserInfoReceiver.ACTION.UPDATE_USERINFO
44                                .equals(action)) {
45                            String type=intent.getStringExtra(
46                            UpdateUserInfoReceiver.INTENT_TYPE.TYPE_NAME);
47                            //更新头像
48                            if(UpdateUserInfoReceiver.INTENT_TYPE.UPDATE_HEAD
49                                    .equals(type)) {
50                                String head=intent.getStringExtra("head");
51                                Bitmap bt=BitmapFactory.decodeFile(head);
52                                if(bt !=null) {
53                                    Drawable drawable=new BitmapDrawable(bt);
54                                    iv_avatar.setImageDrawable(drawable);
55                                } else {
56                                    iv_avatar.setImageResource(R.drawable.default_head);
57                                }
58                            }
59                        }
60                    }
61                });
62        filter=new IntentFilter(UpdateUserInfoReceiver.ACTION.UPDATE_USERINFO);
63        getActivity().registerReceiver(updateUserInfoReceiver, filter);
64    }
65    private void setListener() {
66        ll_calendar.setOnClickListener(this);
67        ll_constellation.setOnClickListener(this);
68        ll_scraw.setOnClickListener(this);
69        ll_map.setOnClickListener(this);
70        rl_collection.setOnClickListener(this);
```

```java
71          rl_setting.setOnClickListener(this);
72          iv_avatar.setOnClickListener(this);
73      }
74      @Override
75      public void onDestroy() {
76          super.onDestroy();
77          if(updateUserInfoReceiver != null) {
78              getActivity().unregisterReceiver(updateUserInfoReceiver);
79          }
80      }
81      @Override
82      public void onClick(View view) {
83          switch(view.getId()) {
84              case R.id.ll_calendar:
85                  break;
86              case R.id.ll_constellation:
87                  break;
88              case R.id.ll_scraw:
89                  break;
90              case R.id.ll_map:
91                  break;
92              case R.id.rl_collection:
93                  if(isLogin) {
94                      //跳转到"收藏"界面
95                  } else {
96                      Toast.makeText(getActivity(), "您还未登录,请先登录",
97                          Toast.LENGTH_SHORT).show();
98                  }
99                  break;
100             case R.id.rl_setting:
101                 if(isLogin) {
102                     //跳转到"设置"界面
103                 } else {
104                     Toast.makeText(getActivity(), "您还未登录,请先登录",
105                         Toast.LENGTH_SHORT).show();
106                 }
107                 break;
108             case R.id.iv_avatar:
109                 if(isLogin) {
110                     //跳转到"个人资料"界面
111                 } else {
112                     //跳转到"登录"界面
113                 }
114                 break;
115         }
116     }
117     /**
118      * 根据登录状态设置"我"界面
119      */
120     public void setLoginParams(boolean isLogin) {
```

```
121        if(isLogin) {
122            String userName=UtilsHelper.readLoginUserName(getActivity());
123            collapsingToolbarLayout.setTitle(userName);
124            String head=DBUtils.getInstance(getActivity()).getUserHead(userName);
125            Bitmap bt=BitmapFactory.decodeFile(head);
126            if(bt !=null) {
127                Drawable drawable=new BitmapDrawable(bt);       //转换成 drawable
128                iv_avatar.setImageDrawable(drawable);
129            } else {
130                iv_avatar.setImageResource(R.drawable.default_head);
131            }
132        } else {
133            iv_avatar.setImageResource(R.drawable.default_head);
134            collapsingToolbarLayout.setTitle("点击登录");
135        }
136    }
137    @Override
138    public void onActivityResult(int requestCode, int resultCode, Intent data) {
139        super.onActivityResult(requestCode, resultCode, data);
140        if(data !=null) {
141            boolean isLogin=data.getBooleanExtra("isLogin", false);
142            setLoginParams(isLogin);
143            this.isLogin=isLogin;
144        }
145    }
146 }
```

(6) 修改 UtilsHelper.java 文件。由于"我"界面需要根据登录状态设置相应的图标和控件的显示，因此需要在 com.itheima.topline.utils 包中的 UtilsHelper 类中创建 readLoginStatus() 方法，从 SharedPreferences 中读取登录状态，具体代码如下：

```
/**
 * 从 SharedPreferences 中读取登录状态
 */
public static boolean readLoginStatus(Context context){
    SharedPreferences sp=context.getSharedPreferences("loginInfo",Context.MODE_PRIVATE);
    boolean isLogin=sp.getBoolean("isLogin", false);
    return isLogin;
}
```

由于在"我"界面中需要显示用户名，因此当用户处于登录状态时，需要在 com.itheima.topline.utils 包中的 UtilsHelper 类中创建 readLoginUserName() 方法，从 SharedPreferences 中读取用户名，具体代码如下：

```
/**
 * 从 SharedPreferences 中读取登录用户名
 */
public static String readLoginUserName(Context context) {
    SharedPreferences sp=context.getSharedPreferences("loginInfo",
```

```
        Context.MODE_PRIVATE);
String userName=sp.getString("loginUserName", "");         //读取登录时的用户名
return userName;
}
```

(7) 修改 DBUtils.java 文件。由于"我"界面中需要根据用户名获取头像,因此需要在 com.itheima.topline.utils 包中的 DBUtils 类中添加如下代码:

```
/*
 * 根据登录名获取用户头像
 */
public String getUserHead(String userName) {
    String sql="SELECT head FROM "+SQLiteHelper.U_USERINFO+" WHERE userName=?";
    Cursor cursor=db.rawQuery(sql, new String[]{userName});
    String head="";
    while (cursor.moveToNext()) {
        head=cursor.getString(cursor.getColumnIndex("head"));
    }
    cursor.close();
    return head;
}
```

(8) 修改 styles.xml 文件。由于在"我"界面中需要设置标题字体的大小,因此需要在 res/values 文件夹的 styles.xml 文件中添加如下代码:

```
<!--toolbar 标题样式 -->
<style name="ToolbarTitle" parent="@style/TextAppearance.Widget.AppCompat.Toolbar.
    Title">
    <item name="android:textSize">18sp</item>
</style>
```

(9) 修改底部导航栏。由于点击底部导航栏的"我"按钮时会出现"我"界面,因此需要找到第 6 章中文件 6-8 中的 initView()方法,在该方法中的"VideoFragment videoFragment =new VideoFragment();"语句下方添加如下代码:

```
MeFragment meFragment=new MeFragment();
```

在"alFragment.add(videoFragment);"语句下方添加如下代码:

```
alFragment.add(meFragment);
viewPager.setOffscreenPageLimit(3);         //三个界面之间来回切换都不会重新加载数据
```

10.3 注册

任务综述

"注册"界面主要用于输入注册信息(用户名与密码),点击"注册"按钮即可进行注册。

由于头条项目把用户信息全部存放在本地,因此在注册成功后,需要将用户名和密码保存在 SharedPreferences 中便于后续用户登录。为了保证账户的安全,在保存密码时会采用 MD5 加密算法,这种算法是不可逆的,且具有一定的安全性。

【任务 10-7】 "注册"界面

【任务分析】

"注册"界面主要用于输入用户的注册信息,实现用户注册功能,界面效果如图 10-2 所示。

【任务实施】

(1)创建"注册"界面。在 com.itheima.topline.activity 包中创建一个 Empty Activity 类,命名为 RegisterActivity 并将布局文件名指定为 activity_register。在该布局文件中,通过＜include＞标签将 main_title_bar.xml(标题栏)引入。

(2)导入界面图片。将"注册"界面所需的图片 user_name_icon.png、psw_icon.png、hide_psw_icon.png、show_psw_icon.png 导入 drawable-hdpi 文件夹。

(3)放置界面控件。在布局文件中,放置一个 ImageViewRoundOval 控件,用于显示用户默认头像;两个 EditText 控件,用于输入用户名和密码;一个

图 10-2 "注册"界面

ImageView 控件,用于密码的显示与隐藏;一个 Button 控件,作为注册按钮。具体代码如文件 10-11 所示。

【文件 10-11】 activity_register.xml

```
1  <?xml version="1.0" encoding="utf-8"?>
2  <LinearLayout xmlns:android="http://schemas.android.com/apk/res/android"
3      xmlns:app="http://schemas.android.com/apk/res-auto"
4      android:layout_width="match_parent"
5      android:layout_height="match_parent"
6      android:background="@color/register_bg_color"
7      android:orientation="vertical">
8      <include layout="@layout/main_title_bar" />
9      <com.itheima.topline.view.ImageViewRoundOval
10         android:layout_width="80dp"
11         android:layout_height="80dp"
12         android:layout_gravity="center_horizontal"
13         android:layout_marginBottom="25dp"
14         android:layout_marginTop="35dp"
15         android:scaleType="fitXY"
16         android:src="@drawable/default_head" />
17     <android.support.v7.widget.CardView
```

```xml
            android:layout_width="match_parent"
            android:layout_height="wrap_content"
            android:layout_marginLeft="25dp"
            android:layout_marginRight="25dp"
            app:cardCornerRadius="5dp"
            app:cardElevation="3dp"
            app:cardPreventCornerOverlap="false"
            app:cardUseCompatPadding="true">
            <LinearLayout
                android:layout_width="match_parent"
                android:layout_height="wrap_content"
                android:orientation="vertical">
                <EditText
                    android:id="@+id/et_user_name"
                    android:layout_width="fill_parent"
                    android:layout_height="48dp"
                    android:layout_gravity="center_horizontal"
                    android:background="@drawable/register_edittext_top_radius"
                    android:drawableLeft="@drawable/user_name_icon"
                    android:drawablePadding="10dp"
                    android:gravity="center_vertical"
                    android:hint="请输入用户名"
                    android:paddingLeft="8dp"
                    android:singleLine="true"
                    android:textColor="#000000"
                    android:textColorHint="@color/register_hint_text_color"
                    android:textCursorDrawable="@null"
                    android:textSize="14sp" />
                <View
                    android:layout_width="fill_parent"
                    android:layout_height="1dp"
                    android:background="@color/divider_line_color" />
                <RelativeLayout
                    android:layout_width="fill_parent"
                    android:layout_height="wrap_content"
                    android:gravity="center_vertical"
                    android:orientation="horizontal">
                    <EditText
                        android:id="@+id/et_psw"
                        android:layout_width="fill_parent"
                        android:layout_height="48dp"
                        android:layout_gravity="center_horizontal"
                        android:background="@drawable/register_edittext_bottom_radius"
                        android:drawableLeft="@drawable/psw_icon"
                        android:drawablePadding="10dp"
                        android:hint="请输入密码"
                        android:inputType="textPassword"
                        android:paddingLeft="8dp"
                        android:singleLine="true"
                        android:textColor="#000000"
```

```xml
68              android:textColorHint="@color/register_hint_text_color"
69              android:textCursorDrawable="@null"
70              android:textSize="14sp" />
71          <ImageView
72              android:id="@+id/iv_show_psw"
73              android:layout_width="15dp"
74              android:layout_height="48dp"
75              android:layout_alignParentRight="true"
76              android:layout_centerVertical="true"
77              android:layout_marginRight="8dp"
78              android:src="@drawable/hide_psw_icon" />
79      </RelativeLayout>
80    </LinearLayout>
81  </android.support.v7.widget.CardView>
82  <Button
83      android:id="@+id/btn_register"
84      android:layout_width="fill_parent"
85      android:layout_height="35dp"
86      android:layout_gravity="center_horizontal"
87      android:layout_marginLeft="25dp"
88      android:layout_marginRight="25dp"
89      android:layout_marginTop="15dp"
90      android:background="@drawable/register_btn_selector"
91      android:text="注 册"
92      android:textColor="@android:color/white"
93      android:textSize="18sp" />
94 </LinearLayout>
```

(4) 自定义 ImageViewRoundOval 控件。"注册"界面中有一个圆形默认头像，该效果是通过自定义 ImageViewRoundOval 控件实现的。在 com.itheima.topline.view 包中创建一个 ImageViewRoundOval 类并继承 ImageView 类，具体代码如文件 10-12 所示。

【文件 10-12】 ImageViewRoundOval.java

```java
1  package com.itheima.topline.view;
2  /**
3   * 实现圆形、圆角、椭圆等自定义图片 View
4   */
5  public class ImageViewRoundOval extends ImageView {
6      private Paint mPaint;
7      private int mWidth;
8      private int mRadius;                                  //圆半径
9      private RectF mRect;                                  //矩形凹行大小
10     private int mRoundRadius;                             //圆角大小
11     private BitmapShader mBitmapShader;                   //图形渲染
12     private Matrix mMatrix;
13     private int mType;                                    //记录是圆形还是圆角矩形
14     public static final int TYPE_CIRCLE=0;                //圆形
15     public static final int TYPE_ROUND=1;                 //圆角矩形
16     public static final int TYPE_OVAL=2;                  //椭圆形
17     public static final int DEFAUT_ROUND_RADIUS=10;       //默认圆角大小
```

```java
18    public ImageViewRoundOval(Context context) {
19        this(context, null);
20    }
21    public ImageViewRoundOval(Context context, AttributeSet attrs) {
22        this(context, attrs, 0);
23    }
24    public ImageViewRoundOval(Context context, AttributeSet attrs, int defStyle) {
25        super(context, attrs, defStyle);
26        initView();
27    }
28    private void initView() {
29        mPaint=new Paint();
30        mPaint.setAntiAlias(true);
31        mMatrix=new Matrix();
32        mRoundRadius=DEFAUT_ROUND_RADIUS;
33    }
34    @Override
35    protected void onMeasure(int widthMeasureSpec, int heightMeasureSpec) {
36        super.onMeasure(widthMeasureSpec, heightMeasureSpec);
37        //如果是绘制圆形,则强制宽和高大小一致
38        if(mType==TYPE_CIRCLE) {
39            mWidth=Math.min(getMeasuredWidth(), getMeasuredHeight());
40            mRadius=mWidth / 2;
41            setMeasuredDimension(mWidth, mWidth);
42        }
43    }
44    @Override
45    protected void onDraw(Canvas canvas) {
46        if(null==getDrawable()) {
47            return;
48        }
49        setBitmapShader();
50        if(mType==TYPE_CIRCLE) {
51            canvas.drawCircle(mRadius, mRadius, mRadius, mPaint);
52        } else if(mType==TYPE_ROUND) {
53            mPaint.setColor(Color.RED);
54            canvas.drawRoundRect(mRect, mRoundRadius, mRoundRadius, mPaint);
55        }else if(mType==TYPE_OVAL){
56            canvas.drawOval(mRect, mPaint);
57        }
58    }
59    @Override
60    protected void onSizeChanged(int w, int h, int oldw, int oldh) {
61        super.onSizeChanged(w, h, oldw, oldh);
62        mRect=new RectF(0, 0, getWidth(), getHeight());
63    }
64    /**
65     * 设置 BitmapShader
66     */
67    private void setBitmapShader() {
```

```java
68          Drawable drawable=getDrawable();
69          if(null==drawable){
70              return;
71          }
72          Bitmap bitmap=drawableToBitmap(drawable);
73          //将bitmap作为着色器,创建一个BitmapShader
74          mBitmapShader=new BitmapShader(bitmap, Shader.TileMode.CLAMP,
75          Shader.TileMode.CLAMP);
76          float scale=1.0f;
77          if(mType==TYPE_CIRCLE) {
78              //得到bitmap宽或高的小值
79              int bSize=Math.min(bitmap.getWidth(), bitmap.getHeight());
80              scale=mWidth * 1.0f / bSize;
81          } else if(mType==TYPE_ROUND || mType==TYPE_OVAL) {
82              //如果图片的宽或高与view的宽和高不匹配,则计算出需要缩放的比例;缩放后的图片
83              //的宽和高,一定要大于view的宽和高,所以这里取大值
84              scale=Math.max(getWidth() * 1.0f / bitmap.getWidth(), getHeight()
85                      * 1.0f /bitmap.getHeight());
86          }
87          //shader的变换矩阵,这里主要用于放大或缩小
88          mMatrix.setScale(scale, scale);
89          //设置变换矩阵
90          mBitmapShader.setLocalMatrix(mMatrix);
91          mPaint.setShader(mBitmapShader);
92      }
93      /**
94       * drawable转bitmap
95       * @param drawable
96       */
97      private Bitmap drawableToBitmap(Drawable drawable) {
98          if(drawable instanceof BitmapDrawable) {
99              BitmapDrawable bitmapDrawable= (BitmapDrawable) drawable;
100             return bitmapDrawable.getBitmap();
101         }
102         int w=drawable.getIntrinsicWidth();
103         int h=drawable.getIntrinsicHeight();
104         Bitmap bitmap=Bitmap.createBitmap(w, h, Bitmap.Config.ARGB_8888);
105         Canvas canvas=new Canvas(bitmap);
106         drawable.setBounds(0, 0, w, h);
107         drawable.draw(canvas);
108         return bitmap;
109     }
110     public int getType() {
111         return mType;
112     }
113     /**
114      * 设置图片类型:圆形、圆角矩形、椭圆形
115      * @param mType
116      */
117     public void setType(int mType) {
```

```
118            if(this.mType !=mType){
119                this.mType=mType;
120                invalidate();
121            }
122        }
123 }
```

(5) 修改 colors.xml 文件。由于在"注册"界面用到了"注册"界面的背景颜色、注册文本颜色以及"注册"按钮的背景颜色,因此需要在 res/values 文件夹的 colors.xml 文件中添加这些颜色值,具体添加代码如下所示:

```
<color name="register_bg_color">#f6f6f6</color>
<color name="register_hint_text_color">#a3a3a3</color>
<color name="register_btn_color">#d6d7d7</color>
```

(6) 设置圆角形状。由于"注册"界面输入用户名与密码的控件需要设置圆角形状,因此需要在 res/drawable 文件夹中创建 register_edittext_top_radius.xml 文件与 register_edittext_bottom_radius.xml 文件,分别用于设置控件的顶部与底部的圆角形状,具体代码如文件 10-13 与文件 10-14 所示。

【文件 10-13】 register_edittext_top_radius.xml

```
1  <?xml version="1.0" encoding="utf-8"?>
2  <layer-list xmlns:android="http://schemas.android.com/apk/res/android">
3      <item>
4          <shape android:shape="rectangle">
5              <solid android:color="@android:color/white" />
6              <corners
7                  android:topLeftRadius="4dp"
8                  android:topRightRadius="4dp" />
9          </shape>
10     </item>
11 </layer-list>
```

【文件 10-14】 register_edittext_bottom_radius.xml

```
1  <?xml version="1.0" encoding="utf-8"?>
2  <layer-list xmlns:android="http://schemas.android.com/apk/res/android">
3      <item>
4          <shape android:shape="rectangle">
5              <solid android:color="@android:color/white" />
6              <corners
7                  android:bottomLeftRadius="4dp"
8                  android:bottomRightRadius="4dp" />
9          </shape>
10     </item>
11 </layer-list>
```

(7) 创建背景选择器。在 res/drawable 文件夹中创建"注册"按钮的背景选择器 register_btn_selector.xml。当按钮按下时显示灰色背景,当按钮弹起时显示红色背景,具体代码如文件 10-15 所示。

【文件 10-15】 register_btn_selector.xml

```xml
1   <?xml version="1.0" encoding="utf-8"?>
2   <selector xmlns:android="http://schemas.android.com/apk/res/android">
3       <item android:state_pressed="true">
4           <shape android:shape="rectangle">
5               <corners android:radius="4dp" />
6               <solid android:color="@color/register_btn_color" />
7           </shape>
8       </item>
9       <item android:state_pressed="false">
10          <shape android:shape="rectangle">
11              <corners android:radius="4dp" />
12              <solid android:color="@color/rdTextColorPress" />
13          </shape>
14      </item>
15  </selector>
```

【任务 10-8】 MD5 加密算法

【任务分析】

MD5 的全称是 Message-Digest Algorithm 5（信息-摘要算法），MD5 算法简单来说就是把任意长度的字符串变换成固定长度（通常是 128 位）的十六进制字符串。在存储密码过程中，直接存储明文密码是很危险的，因此在存储密码前需要使用 MD5 算法加密，这样不仅提高了用户信息的安全性，同时也增加了密码破解的难度。

【任务实施】

（1）创建 MD5Utils 类。在 com.itheima.topline.utils 包中创建一个 Java 类，命名为 MD5Utils。

（2）进行 MD5 加密。在 MD5Utils 类中，创建一个 md5() 方法对密码进行加密。首先通过 MessageDigest 的 getInstance() 方法获取数据加密对象 digest，然后通过该对象的 digest() 方法对密码进行加密，具体代码如文件 10-16 所示。

【文件 10-16】 MD5Utils.java

```java
1   package com.itheima.topline.utils;
2   public class MD5Utils {
3       /**
4        * MD5 加密的算法
5        */
6       public static String md5(String text) {
7           MessageDigest digest=null;
8           try {
9               digest=MessageDigest.getInstance("md5");
10              byte[] result=digest.digest(text.getBytes());
```

```
11              StringBuilder sb=new StringBuilder();
12              for(byte b : result) {
13                  int number=b & 0xff;
14                  String hex=Integer.toHexString(number);
15                  if(hex.length()==1) {
16                      sb.append("0"+hex);
17                  } else {
18                      sb.append(hex);
19                  }
20              }
21              return sb.toString();
22          } catch(NoSuchAlgorithmException e) {
23              e.printStackTrace();
24              return "";
25          }
26      }
27  }
```

【任务 10-9】 "注册"界面逻辑代码

【任务分析】

在"注册"界面点击"注册"按钮后,需要获取用户名与密码,将用户名和密码(经过 MD5 加密)保存到 SharedPreferences 中。同时,注册成功后需要将用户名传递到"登录"界面(LoginActivity 目前还未创建)中。

【任务实施】

(1) 获取界面控件。在 RegisterActivity 中创建界面控件的初始化方法 init(),用于获取"注册"界面所要用到的控件以及实现控件的点击事件。

(2) 判断 SharedPreferences 中是否存在需要注册的用户名。在 RegisterActivity 中创建一个 isExistUserName()方法,用于判断 SharedPreferences 中是否存在需要注册的用户名。

(3) 保存注册信息。在 RegisterActivity 中创建一个 saveRegisterInfo()方法,将注册成功的用户名和密码(经过 MD5 加密)保存到 SharedPreferences 中,具体代码如文件 10-17 所示。

【文件 10-17】 RegisterActivity.java

```
1  package com.itheima.topline.activity;
2  public class RegisterActivity extends AppCompatActivity implements View.
3  OnClickListener {
4      private TextView tv_main_title, tv_back;
5      private RelativeLayout rl_title_bar;
6      private SwipeBackLayout layout;
7      private EditText et_psw, et_user_name;
8      private ImageView iv_show_psw;
```

```java
9       private Button btn_register;
10      private String userName, psw;
11      private boolean isShowPsw=false;
12      @Override
13      protected void onCreate(Bundle savedInstanceState) {
14          super.onCreate(savedInstanceState);
15          layout=(SwipeBackLayout) LayoutInflater.from(this).inflate(
16                  R.layout.base, null);
17          layout.attachToActivity(this);
18          setContentView(R.layout.activity_register);
19          init();
20      }
21      private void init() {
22          tv_main_title=(TextView) findViewById(R.id.tv_main_title);
23          tv_main_title.setText("注册");
24          rl_title_bar=(RelativeLayout) findViewById(R.id.title_bar);
25          rl_title_bar.setBackgroundColor(getResources().getColor(R.color.
26                  rdTextColorPress));
27          tv_back=(TextView) findViewById(R.id.tv_back);
28          tv_back.setVisibility(View.VISIBLE);
29          btn_register=(Button) findViewById(R.id.btn_register);
30          et_user_name=(EditText) findViewById(R.id.et_user_name);
31          et_psw=(EditText) findViewById(R.id.et_psw);
32          iv_show_psw=(ImageView) findViewById(R.id.iv_show_psw);
33          tv_back.setOnClickListener(this);
34          iv_show_psw.setOnClickListener(this);
35          btn_register.setOnClickListener(this);
36      }
37      @Override
38      public void onClick(View view) {
39          switch(view.getId()) {
40              case R.id.tv_back:
41                  RegisterActivity.this.finish();
42                  break;
43              case R.id.iv_show_psw:
44                  psw=et_psw.getText().toString();
45                  if(isShowPsw) {
46                      iv_show_psw.setImageResource(R.drawable.hide_psw_icon);
47                      et_psw.setTransformationMethod(PasswordTransformationMethod.
48                                          getInstance());        //隐藏密码
49                      isShowPsw=false;
50                      if(psw !=null) {
51                          et_psw.setSelection(psw.length());
52                      }
53                  } else {
54                      iv_show_psw.setImageResource(R.drawable.show_psw_icon);
55                      et_psw.setTransformationMethod(HideReturnsTransformationMethod.
56                                          getInstance());        //显示密码
57                      isShowPsw=true;
58                      if(psw !=null) {
```

```
59                      et_psw.setSelection(psw.length());
60                  }
61              }
62              break;
63          case R.id.btn_register:
64              //获取输入在相应控件中的字符串
65              userName=et_user_name.getText().toString().trim();
66              psw=et_psw.getText().toString().trim();
67              if(TextUtils.isEmpty(userName)) {
68                  Toast.makeText(RegisterActivity.this, "请输入用户名",
69                          Toast.LENGTH_SHORT).show();
70                  return;
71              } else if(TextUtils.isEmpty(psw)) {
72                  Toast.makeText(RegisterActivity.this, "请输入密码",
73                          Toast.LENGTH_SHORT).show();
74                  return;
75              } else if(isExistUserName(userName)) {
76                  Toast.makeText(RegisterActivity.this, "此账户名已经存在",
77                          Toast.LENGTH_SHORT).show();
78                  return;
79              } else {
80                  Toast.makeText(RegisterActivity.this, "注册成功",
81                          Toast.LENGTH_SHORT).show();
82                  //把用户名和密码保存到SharedPreferences中
83                  saveRegisterInfo(userName, psw);
84                  //注册成功后把用户名传递到LoginActivity.java中
85                  Intent data=new Intent();
86                  data.putExtra("userName", userName);
87                  setResult(RESULT_OK, data);
88                  RegisterActivity.this.finish();
89              }
90              break;
91          }
92      }
93      /**
94       * 从SharedPreferences中读取输入的用户名,判断SharedPreferences中是否有此用户名
95       */
96      private boolean isExistUserName(String userName) {
97          boolean has_userName=false;
98          SharedPreferences sp=getSharedPreferences("loginInfo", MODE_PRIVATE);
99          String spPsw=sp.getString(userName, "");
100         if(!TextUtils.isEmpty(spPsw)) {
101             has_userName=true;
102         }
103         return has_userName;
104     }
105     /**
106      * 保存用户名和密码到SharedPreferences中
107      */
108     private void saveRegisterInfo(String userName, String psw) {
```

```
109        String md5Psw=MD5Utils.md5(psw);              //把密码用 MD5 加密
110        //loginInfo 表示文件名
111        SharedPreferences sp=getSharedPreferences("loginInfo", MODE_PRIVATE);
112        SharedPreferences.Editor editor=sp.edit();     //获取编辑器
113        //以用户名为 key、密码为 value 保存到 SharedPreferences 中
114        editor.putString(userName, md5Psw);
115        editor.commit();             //提交修改
116    }
117 }
```

（4）修改清单文件。由于注册界面向右滑动会关闭该界面，因此需要给该界面添加透明主题的样式，在清单文件的 RegisterActivity 对应的 activity 标签中添加如下代码：

```
android:theme="@style/AppTheme.TransparentActivity"
```

10.4 登录

任务综述

"登录"界面主要是为用户提供一个输入登录信息的界面，当点击"登录"按钮时，需要在 SharedPreferences 中查询输入的用户名是否有对应的密码，如果有，则用此密码与当前输入的密码（需 MD5 加密）进行比对，如果信息一致，则登录成功，并把登录成功的状态和用户名保存到 SharedPreferences 中，便于后续判断登录状态和获取用户名。如果登录失败，则有两种情况，一种是输入的用户名和密码不一致，另一种是此用户名不存在。

【任务 10-10】 "登录"界面

【任务分析】

"登录"界面主要是为用户提供一个登录的入口，若用户还未注册，则可以点击"快速注册"按钮进入注册界面；若用户忘记密码，则可以点击"忘记密码？"按钮进入"找回密码"界面（"找回密码"界面暂时未创建）。界面效果如图 10-3 所示。

【任务实施】

（1）创建"登录"界面。在 com.itheima.topline. activity 包中创建一个 Empty Activity 类，命名为 LoginActivity 并将布局文件名指定为 activity_login。在该布局文件中，通过＜include＞标签将 main_title_ bar.xml（标题栏）引入。

（2）放置界面控件。在布局文件中，放置一个 ImageViewRoundOval 控件，用于显示用户默认头像；

图 10-3 "登录"界面

两个 EditText 控件,分别用于输入用户名和密码;一个 Button 控件,作为登录按钮(和注册按钮使用同一个背景选择器);两个 TextView 控件,分别用于显示文字"快速注册"和"忘记密码?"。具体代码如文件 10-18 所示。

【文件 10-18】 activity_login.xml

```
1   <?xml version="1.0" encoding="utf-8"?>
2   <LinearLayout xmlns:android="http://schemas.android.com/apk/res/android"
3       xmlns:app="http://schemas.android.com/apk/res-auto"
4       android:layout_width="match_parent"
5       android:layout_height="match_parent"
6       android:background="@color/register_bg_color"
7       android:orientation="vertical">
8       <include layout="@layout/main_title_bar" />
9       <com.itheima.topline.view.ImageViewRoundOval
10          android:layout_width="80dp"
11          android:layout_height="80dp"
12          android:layout_gravity="center_horizontal"
13          android:layout_marginBottom="25dp"
14          android:layout_marginTop="35dp"
15          android:scaleType="fitXY"
16          android:src="@drawable/default_head" />
17      <android.support.v7.widget.CardView
18          android:layout_width="match_parent"
19          android:layout_height="wrap_content"
20          android:layout_marginLeft="25dp"
21          android:layout_marginRight="25dp"
22          app:cardCornerRadius="5dp"
23          app:cardElevation="3dp"
24          app:cardPreventCornerOverlap="false"
25          app:cardUseCompatPadding="true">
26          <LinearLayout
27              android:layout_width="match_parent"
28              android:layout_height="wrap_content"
29              android:orientation="vertical">
30              <EditText
31                  android:id="@+id/et_user_name"
32                  android:layout_width="fill_parent"
33                  android:layout_height="48dp"
34                  android:layout_gravity="center_horizontal"
35                  android:background="@drawable/register_edittext_top_radius"
36                  android:drawableLeft="@drawable/user_name_icon"
37                  android:drawablePadding="10dp"
38                  android:gravity="center_vertical"
39                  android:hint="请输入用户名"
40                  android:paddingLeft="8dp"
41                  android:singleLine="true"
42                  android:textColor="#000000"
43                  android:textColorHint="@color/register_hint_text_color"
44                  android:textCursorDrawable="@null"
```

```xml
45              android:textSize="14sp" />
46          <View
47              android:layout_width="fill_parent"
48              android:layout_height="1dp"
49              android:background="@color/divider_line_color" />
50          <RelativeLayout
51              android:layout_width="fill_parent"
52              android:layout_height="48dp"
53              android:gravity="center_vertical"
54              android:orientation="horizontal">
55              <EditText
56                  android:id="@+id/et_psw"
57                  android:layout_width="fill_parent"
58                  android:layout_height="fill_parent"
59                  android:layout_gravity="center_horizontal"
60                  android:background="@drawable/register_edittext_bottom_radius"
61                  android:drawableLeft="@drawable/psw_icon"
62                  android:drawablePadding="10dp"
63                  android:hint="请输入密码"
64                  android:inputType="textPassword"
65                  android:paddingLeft="8dp"
66                  android:singleLine="true"
67                  android:textColor="#000000"
68                  android:textColorHint="@color/register_hint_text_color"
69                  android:textCursorDrawable="@null"
70                  android:textSize="14sp" />
71              <ImageView
72                  android:id="@+id/iv_show_psw"
73                  android:layout_width="15dp"
74                  android:layout_height="fill_parent"
75                  android:layout_alignParentRight="true"
76                  android:layout_centerVertical="true"
77                  android:layout_marginRight="8dp"
78                  android:src="@drawable/hide_psw_icon" />
79          </RelativeLayout>
80      </LinearLayout>
81  </android.support.v7.widget.CardView>
82  <LinearLayout
83      android:layout_width="wrap_content"
84      android:layout_height="wrap_content"
85      android:layout_gravity="right"
86      android:layout_marginLeft="25dp"
87      android:layout_marginRight="25dp"
88      android:orientation="horizontal">
89      <TextView
90          android:id="@+id/tv_quick_register"
91          android:layout_width="wrap_content"
92          android:layout_height="wrap_content"
93          android:text="快速注册"
94          android:textColor="@color/register_hint_text_color"
```

```
 95              android:textSize="12sp" />
 96          <TextView
 97              android:id="@+id/tv_forget_psw"
 98              android:layout_width="wrap_content"
 99              android:layout_height="wrap_content"
100              android:layout_marginLeft="8dp"
101              android:text="忘记密码?"
102              android:textColor="@color/register_hint_text_color"
103              android:textSize="12sp" />
104      </LinearLayout>
105      <Button
106          android:id="@+id/btn_login"
107          android:layout_width="fill_parent"
108          android:layout_height="35dp"
109          android:layout_gravity="center_horizontal"
110          android:layout_marginLeft="25dp"
111          android:layout_marginRight="25dp"
112          android:layout_marginTop="15dp"
113          android:background="@drawable/register_btn_selector"
114          android:text="登 录"
115          android:textColor="@android:color/white"
116          android:textSize="18sp" />
117  </LinearLayout>
```

【任务 10-11】 "登录"界面逻辑代码

【任务分析】

当点击"登录"按钮时,需要先判断用户名和密码是否为空,若为空,则提示"请输入用户名和密码";若不为空,则获取用户输入的用户名。由于头条项目登录、注册数据使用的是本地数据,因此需要根据用户名在 SharedPreferences 中查询是否有对应的密码,如果有对应的密码且密码与用户输入的密码(需 MD5 加密)比对一致,则登录成功。

【任务实施】

(1) 获取界面控件。在 LoginActivity 中创建界面控件的初始化方法 init(),用于获取"登录"界面所要用的控件并设置登录按钮、返回键、快速注册、忘记密码的点击事件。

(2) 获取回传数据。在 LoginActivity 中重写 onActivityResult()方法,通过 data.getStringExtra()方法获取注册成功的用户名,并将其显示在用户名控件上。

(3) 保存登录状态到 SharedPreferences 中。由于在后续创建"我"界面时,需要根据登录状态设置界面的图标和用户名,因此需要创建 saveLoginStatus()方法,在登录成功时把登录状态和用户名保存到 SharedPreferences 中,具体代码如文件 10-19 所示。

【文件 10-19】 LoginActivity.java

```
1  package com.itheima.topline.activity;
2  public class LoginActivity extends AppCompatActivity implements View.
3  OnClickListener{
```

```java
4       private EditText et_psw,et_user_name;
5       private TextView tv_quick_register,tv_forget_psw;
6       private ImageView iv_show_psw;
7       private Button btn_login;
8       private boolean isShowPsw=false;
9       private String userName,psw,spPsw;
10      private TextView tv_main_title,tv_back;
11      private RelativeLayout rl_title_bar;
12      private SwipeBackLayout layout;
13      @Override
14      protected void onCreate(Bundle savedInstanceState) {
15          super.onCreate(savedInstanceState);
16          layout=(SwipeBackLayout) LayoutInflater.from(this).inflate(
17          R.layout.base, null);
18          layout.attachToActivity(this);
19          setContentView(R.layout.activity_login);
20          init();
21      }
22      private void init(){
23          tv_main_title=(TextView)findViewById(R.id.tv_main_title);
24          tv_main_title.setText("登录");
25          rl_title_bar=(RelativeLayout)findViewById(R.id.title_bar);
26          rl_title_bar.setBackgroundColor(getResources().getColor(R.color.
27          rdTextColorPress));
28          tv_back= (TextView) findViewById(R.id.tv_back);
29          tv_back.setVisibility(View.VISIBLE);
30          et_user_name=(EditText) findViewById(R.id.et_user_name);
31          et_psw=(EditText) findViewById(R.id.et_psw);
32          iv_show_psw=(ImageView) findViewById(R.id.iv_show_psw);
33          tv_quick_register=(TextView) findViewById(R.id.tv_quick_register);
34          tv_forget_psw=(TextView) findViewById(R.id.tv_forget_psw);
35          btn_login=(Button) findViewById(R.id.btn_login);
36          tv_back.setOnClickListener(this);
37          iv_show_psw.setOnClickListener(this);
38          tv_quick_register.setOnClickListener(this);
39          tv_forget_psw.setOnClickListener(this);
40          btn_login.setOnClickListener(this);
41      }
42      @Override
43      public void onClick(View view) {
44        switch(view.getId()){
45          case R.id.tv_back:
46              LoginActivity.this.finish();
47              break;
48          case R.id.iv_show_psw:
49              psw=et_psw.getText().toString();
50              if(isShowPsw){
51                  iv_show_psw.setImageResource(R.drawable.hide_psw_icon);
52                  //隐藏密码
53                  et_psw.setTransformationMethod(PasswordTransformationMethod.
```

```java
54              getInstance();
55          isShowPsw=false;
56          if(psw!=null){
57              et_psw.setSelection(psw.length());
58          }
59      }else{
60          iv_show_psw.setImageResource(R.drawable.show_psw_icon);
61          //显示密码
62          et_psw.setTransformationMethod(HideReturnsTransformationMethod.
63              getInstance());
64          isShowPsw=true;
65          if(psw!=null){
66              et_psw.setSelection(psw.length());
67          }
68      }
69      break;
70  case R.id.btn_login:
71      userName=et_user_name.getText().toString().trim();
72      psw=et_psw.getText().toString().trim();
73      String md5Psw=MD5Utils.md5(psw);
74      spPsw=readPsw(userName);
75      if(TextUtils.isEmpty(userName)){
76          Toast.makeText(LoginActivity.this,"请输入用户名",
77              Toast.LENGTH_SHORT).show();
78          return;
79      }else if(TextUtils.isEmpty(psw)){
80          Toast.makeText(LoginActivity.this,"请输入密码",
81              Toast.LENGTH_SHORT).show();
82          return;
83      }else if(md5Psw.equals(spPsw)){
84          Toast.makeText(LoginActivity.this,"登录成功",
85              Toast.LENGTH_SHORT).show();
86          //保存登录状态和登录的用户名
87          saveLoginStatus(true,userName);
88          //把登录成功的状态传递到MeFragment中
89          Intent data=new Intent();
90          data.putExtra("isLogin", true);
91          setResult(RESULT_OK, data);
92          LoginActivity.this.finish();
93          return;
94      }else if((!TextUtils.isEmpty(spPsw)&&!md5Psw.equals(spPsw))){
95          Toast.makeText(LoginActivity.this,"输入的用户名和密码不一致",
96              Toast.LENGTH_SHORT).show();
97          return;
98      }else{
99          Toast.makeText(LoginActivity.this,"此用户名不存在",
100             Toast.LENGTH_SHORT).show();
101     }
102     break;
103 case R.id.tv_quick_register:
```

```java
104                Intent intent=new Intent(LoginActivity.this,RegisterActivity.class);
105                startActivityForResult(intent,1);
106                break;
107            case R.id.tv_forget_psw:
108                break;
109        }
110    }
111    /**
112     * 从SharedPreferences中根据用户名读取密码
113     */
114    private String readPsw(String userName){
115        SharedPreferences sp=getSharedPreferences("loginInfo", MODE_PRIVATE);
116        return sp.getString(userName, "");
117    }
118    /**
119     * 保存登录状态和登录用户名到SharedPreferences中
120     */
121    private void saveLoginStatus(boolean status,String userName){
122        //loginInfo 表示文件名
123        SharedPreferences sp=getSharedPreferences("loginInfo", MODE_PRIVATE);
124        SharedPreferences.Editor editor=sp.edit();              //获取编辑器
125        editor.putBoolean("isLogin", status);                   //存入boolean类型的登录状态
126        editor.putString("loginUserName", userName);            //存入登录时的用户名
127        editor.commit();                                        //提交修改
128    }
129    @Override
130    protected void onActivityResult(int requestCode, int resultCode, Intent data) {
131        super.onActivityResult(requestCode, resultCode, data);
132        if(data!=null){
133            //从"注册"界面传递过来的用户名
134            String userName=data.getStringExtra("userName");
135            if(!TextUtils.isEmpty(userName)){
136                et_user_name.setText(userName);
137                //设置光标的位置
138                et_user_name.setSelection(userName.length());
139            }
140        }
141    }
142 }
```

（4）修改清单文件。由于"登录"界面向右滑动会关闭该界面，因此需要给该界面添加透明主题的样式，在清单文件的LoginActivity对应的activity标签中添加如下代码：

```
android:theme="@style/AppTheme.TransparentActivity"
```

（5）修改MeFragment.java文件。当用户未登录时，点击"我"界面中的头像会跳转到"登录"界面，因此需要找到文件10-10中的onClick()方法，在该方法的注释"//跳转到'登录'界面"下方添加如下代码：

```
Intent login=new Intent(getActivity(), LoginActivity.class);
startActivityForResult(login, 1);
```

10.5 个人资料

任务综述

"个人资料"界面主要用于显示用户信息,其中包含用户头像、用户名、昵称、性别和签名,除了用户名无须修改以外,其余信息均可修改。当注册一个新用户并第一次进入"个人资料"界面时,除用户名以外的信息均使用默认值,当修改个人资料信息时需要使用 SQLite 数据库进行保存。

【任务 10-12】 "个人资料"界面

【任务分析】

"个人资料"界面主要用于展示用户的个人信息,包括用户头像、用户名、昵称、性别和签名,界面效果如图 10-4 所示。

图 10-4 "个人资料"界面

【任务实施】

(1) 创建"个人资料"界面。在 com.itheima.topline.activity 包中创建一个 Empty Activity 类,命名为 UserInfoActivity 并将布局文件名指定为 activity_user_info。在该布局文件中,通过＜include＞标签将 main_title_bar.xml(标题栏)引入。

(2) 放置界面控件。在布局文件中,放置一个 ImageViewRoundOval 控件显示头像;5 个 TextView 控件显示每行标题(头像、用户名、昵称、性别、签名);4 个 TextView 控件显示对应的属性值;5 个 View 控件显示 5 条灰色分隔线,具体代码如文件 10-20 所示。

【文件 10-20】 activity_user_info.xml

```
1    <?xml version="1.0" encoding="utf-8"?>
2    <LinearLayout xmlns:android="http://schemas.android.com/apk/res/android"
3        android:layout_width="match_parent"
4        android:layout_height="match_parent"
5        android:background="@android:color/white"
6        android:orientation="vertical">
7        <include layout="@layout/main_title_bar" />
8        <RelativeLayout
9            android:id="@+id/rl_head"
10           android:layout_width="fill_parent"
11           android:layout_height="60dp"
```

```xml
12          android:layout_marginLeft="15dp"
13          android:layout_marginRight="15dp">
14          <TextView
15              android:layout_width="wrap_content"
16              android:layout_height="wrap_content"
17              android:layout_centerVertical="true"
18              android:text="头    像"
19              android:textColor="#000000"
20              android:textSize="16sp" />
21          <com.itheima.topline.view.ImageViewRoundOval
22              android:id="@+id/iv_head_icon"
23              android:layout_width="40dp"
24              android:layout_height="40dp"
25              android:layout_alignParentRight="true"
26              android:layout_centerVertical="true"
27              android:scaleType="fitXY"
28              android:src="@drawable/default_head" />
29      </RelativeLayout>
30      <View
31          android:layout_width="fill_parent"
32          android:layout_height="1dp"
33          android:background="#E4E4E4" />
34      <RelativeLayout
35          android:id="@+id/rl_account"
36          android:layout_width="fill_parent"
37          android:layout_height="60dp"
38          android:layout_marginLeft="15dp"
39          android:layout_marginRight="15dp">
40          <TextView
41              android:layout_width="wrap_content"
42              android:layout_height="wrap_content"
43              android:layout_centerVertical="true"
44              android:text="用户名"
45              android:textColor="#000000"
46              android:textSize="16sp" />
47          <TextView
48              android:id="@+id/tv_user_name"
49              android:layout_width="wrap_content"
50              android:layout_height="wrap_content"
51              android:layout_alignParentRight="true"
52              android:layout_centerVertical="true"
53              android:layout_marginRight="5dp"
54              android:textColor="#a3a3a3"
55              android:textSize="14sp" />
56      </RelativeLayout>
57      <View
58          android:layout_width="fill_parent"
59          android:layout_height="1dp"
60          android:background="#E4E4E4" />
61      <RelativeLayout
```

```
62          android:id="@+id/rl_nickName"
63          android:layout_width="fill_parent"
64          android:layout_height="60dp"
65          android:layout_marginLeft="15dp"
66          android:layout_marginRight="15dp">
67          <TextView
68              android:layout_width="wrap_content"
69              android:layout_height="wrap_content"
70              android:layout_centerVertical="true"
71              android:text="昵    称"
72              android:textColor="#000000"
73              android:textSize="16sp" />
74          <TextView
75              android:id="@+id/tv_nickName"
76              android:layout_width="wrap_content"
77              android:layout_height="wrap_content"
78              android:layout_alignParentRight="true"
79              android:layout_centerVertical="true"
80              android:layout_marginRight="5dp"
81              android:singleLine="true"
82              android:textColor="#a3a3a3"
83              android:textSize="14sp" />
84      </RelativeLayout>
85      <View
86          android:layout_width="fill_parent"
87          android:layout_height="1dp"
88          android:background="#E4E4E4" />
89      <RelativeLayout
90          android:id="@+id/rl_sex"
91          android:layout_width="fill_parent"
92          android:layout_height="60dp"
93          android:layout_marginLeft="15dp"
94          android:layout_marginRight="15dp">
95          <TextView
96              android:layout_width="wrap_content"
97              android:layout_height="wrap_content"
98              android:layout_centerVertical="true"
99              android:text="性    别"
100             android:textColor="#000000"
101             android:textSize="16sp" />
102         <TextView
103             android:id="@+id/tv_sex"
104             android:layout_width="wrap_content"
105             android:layout_height="wrap_content"
106             android:layout_alignParentRight="true"
107             android:layout_centerVertical="true"
108             android:layout_marginRight="5dp"
109             android:textColor="#a3a3a3"
110             android:textSize="14sp" />
111     </RelativeLayout>
```

```
112        <View
113            android:layout_width="fill_parent"
114            android:layout_height="1dp"
115            android:background="#E4E4E4" />
116        <RelativeLayout
117            android:id="@+id/rl_signature"
118            android:layout_width="fill_parent"
119            android:layout_height="60dp"
120            android:layout_marginLeft="15dp"
121            android:layout_marginRight="15dp">
122            <TextView
123                android:layout_width="wrap_content"
124                android:layout_height="wrap_content"
125                android:layout_centerVertical="true"
126                android:singleLine="true"
127                android:text="签    名"
128                android:textColor="#000000"
129                android:textSize="16sp" />
130            <TextView
131                android:id="@+id/tv_signature"
132                android:layout_width="wrap_content"
133                android:layout_height="wrap_content"
134                android:layout_alignParentRight="true"
135                android:layout_centerVertical="true"
136                android:layout_marginRight="5dp"
137                android:textColor="#a3a3a3"
138                android:textSize="14sp" />
139        </RelativeLayout>
140        <View
141            android:layout_width="fill_parent"
142            android:layout_height="1dp"
143            android:background="#E4E4E4" />
144 </LinearLayout>
```

（3）放置 dialog_select_photo.xml 文件中的控件。由于点击"个人资料"界面中的头像条目时会弹出一个对话框，对话框中有"从相册中选取"与"拍照"两个按钮，因此需要在 res/layout 文件夹中创建一个 dialog_select_photo.xml 文件，在该文件中放置两个 TextView 控件，分别用于显示"从相册中选取"与"拍照"文本；一个 View 控件用于显示灰色分隔线，具体代码如文件 10-21 所示。

【文件 10-21】 dialog_select_photo.xml

```
1  <?xml version="1.0" encoding="utf-8"?>
2  <LinearLayout xmlns:android="http://schemas.android.com/apk/res/android"
3      android:layout_width="match_parent"
4      android:layout_height="match_parent"
5      android:background="@android:color/white"
6      android:gravity="center"
7      android:orientation="vertical">
8      <TextView
```

```
9            android:id="@+id/tv_select_gallery"
10           android:layout_width="match_parent"
11           android:layout_height="50dp"
12           android:gravity="center"
13           android:text="从相册中选取"
14           android:textColor="@android:color/black"
15           android:textSize="14sp" />
16      <View
17           android:layout_width="match_parent"
18           android:layout_height="1dp"
19           android:background="@color/divider_line_color" />
20      <TextView
21           android:id="@+id/tv_select_camera"
22           android:layout_width="match_parent"
23           android:layout_height="50dp"
24           android:gravity="center"
25           android:text="拍照"
26           android:textColor="@android:color/black"
27           android:textSize="14sp" />
28  </LinearLayout>
```

【任务10-13】 "个人资料"界面逻辑代码

【任务分析】

"个人资料"界面主要用于展示用户的相关信息,当进入"个人资料"界面时,首先查询数据库中的用户信息,并将信息展示到界面中。"个人资料"界面中的昵称、性别和签名是可以修改的,因此需要添加相应的监听事件,当点击"昵称"时跳转到"昵称修改"界面,当点击"性别"时弹出"性别选择"对话框,当点击"签名"时跳转到"签名修改"界面。

【任务实施】

（1）获取界面控件。在 UserInfoActivity 中创建界面控件的初始化方法 init(),获取"个人资料"界面所要用到的控件。

（2）设置点击事件。由于界面中除了用户名和头像以外的其余属性值都可以修改,因此需要为其余属性所在的条目设置点击事件。在 UserInfoActivity 类中实现 OnClickListener 接口,然后创建 setListener()方法,在该方法中设置昵称、性别、签名的点击监听事件并实现 OnClickListener 接口中的 onClick()方法。

（3）为界面控件设置值。创建一个 initData()方法用于从数据库中获取数据,如果数据库中的数据为空,则为此账号设置默认的属性值并保存到数据库中,具体代码如文件 10-22 所示。

【文件 10-22】 UserInfoActivity.java

```
1  package com.itheima.topline.activity;
2  public class UserInfoActivity extends AppCompatActivity implements View.
3  OnClickListener{
```

```java
4       private TextView tv_main_title,tv_back;
5       private SwipeBackLayout layout;
6       private TextView tv_nickName, tv_signature, tv_user_name, tv_sex;
7       private RelativeLayout rl_nickName, rl_sex, rl_signature,rl_head, rl_title_bar;
8       private String spUserName;
9       private static final int CROP_PHOTO1=3;              //裁剪图片
10      private static final int CROP_PHOTO2=4;              //裁剪图片
11      private static final int SAVE_PHOTO=5;               //保存图片
12      private ImageViewRoundOval iv_photo;
13      private Bitmap head;                                 //头像 Bitmap
14      private static String path="/sdcard/TopLine/myHead/";       //sd 路径
15      @Override
16      protected void onCreate(Bundle savedInstanceState) {
17          super.onCreate(savedInstanceState);
18          layout= (SwipeBackLayout) LayoutInflater.from(this).inflate(
19                                              R.layout.base, null);
20          layout.attachToActivity(this);
21          setContentView(R.layout.activity_user_info);
22          //从 SharedPreferences 中获取登录时的用户名
23          spUserName=UtilsHelper.readLoginUserName(this);
24          init();
25          initData();
26          setListener();
27      }
28      private void init(){
29          tv_main_title=(TextView)findViewById(R.id.tv_main_title);
30          tv_main_title.setText("个人资料");
31          rl_title_bar=(RelativeLayout)findViewById(R.id.title_bar);
32          rl_title_bar.setBackgroundColor(getResources().getColor(R.color.
33                                              rdTextColorPress));
34          tv_back=(TextView) findViewById(R.id.tv_back);
35          tv_back.setVisibility(View.VISIBLE);
36          rl_nickName=(RelativeLayout) findViewById(R.id.rl_nickName);
37          rl_sex=(RelativeLayout) findViewById(R.id.rl_sex);
38          rl_signature=(RelativeLayout) findViewById(R.id.rl_signature);
39          tv_nickName=(TextView) findViewById(R.id.tv_nickName);
40          tv_user_name=(TextView) findViewById(R.id.tv_user_name);
41          tv_sex=(TextView) findViewById(R.id.tv_sex);
42          tv_signature=(TextView) findViewById(R.id.tv_signature);
43          rl_head=(RelativeLayout) findViewById(R.id.rl_head);
44          iv_photo=(ImageViewRoundOval) findViewById(R.id.iv_head_icon);
45      }
46      /**
47       * 获取数据
48       */
49      private void initData() {
50          UserBean bean=null;
51          bean=DBUtils.getInstance(this).getUserInfo(spUserName);
52          //首先判断数据库是否有数据
53          if(bean==null){
```

```java
54              bean=new UserBean();
55              bean.setUserName(spUserName);
56              bean.setNickName("问答精灵");
57              bean.setSex("男");
58              bean.setSignature("传智播客问答精灵");
59              iv_photo.setImageResource(R.drawable.default_head);
60              //保存用户信息到数据库
61              DBUtils.getInstance(this).saveUserInfo(bean);
62          }
63          setValue(bean);
64      }
65      /**
66       * 为界面控件设置值
67       */
68      private void setValue(UserBean bean) {
69          tv_nickName.setText(bean.getNickName());
70          tv_user_name.setText(bean.getUserName());
71          tv_sex.setText(bean.getSex());
72          tv_signature.setText(bean.getSignature());
73          //从SD卡中寻找头像,转换成Bitmap
74          Bitmap bt=BitmapFactory.decodeFile(bean.getHead());
75          if(bt !=null) {
76              @SuppressWarnings("deprecation")
77              Drawable drawable=new BitmapDrawable(bt);         //转换成drawable
78              iv_photo.setImageDrawable(drawable);
79          } else {
80              iv_photo.setImageResource(R.drawable.default_head);
81          }
82      }
83      /**
84       * 设置控件的点击监听事件
85       */
86      private void setListener() {
87          tv_back.setOnClickListener(this);
88          rl_nickName.setOnClickListener(this);
89          rl_sex.setOnClickListener(this);
90          rl_signature.setOnClickListener(this);
91          rl_head.setOnClickListener(this);
92      }
93      /**
94       * 控件的点击事件
95       */
96      @Override
97      public void onClick(View v) {
98          switch(v.getId()) {
99              case R.id.tv_back:              //返回键的点击事件
100                 this.finish();
101                 break;
102             case R.id.rl_nickName:          //昵称的点击事件
103                 break;
```

```java
104            case R.id.rl_sex:                        //性别的点击事件
105                String sex=tv_sex.getText().toString();   //获取性别控件中的数据
106                sexDialog(sex);
107                break;
108            case R.id.rl_signature:                  //签名的点击事件
109                break;
110            case R.id.rl_head:                       //头像的点击事件
111                showTypeDialog();
112                break;
113            default:
114                break;
115        }
116    }
117    private void showTypeDialog() {
118        AlertDialog.Builder builder=new AlertDialog.Builder(this);
119        final AlertDialog dialog=builder.create();
120        View view=View.inflate(this, R.layout.dialog_select_photo, null);
121        TextView tv_select_gallery= (TextView) view.findViewById(
122                        R.id.tv_select_gallery);
123        TextView tv_select_camera= (TextView) view.findViewById(R.id.tv_select_camera);
124        //在相册中选取
125        tv_select_gallery.setOnClickListener(new View.OnClickListener() {
126            @Override
127            public void onClick(View v) {
128                Intent intent1=new Intent(Intent.ACTION_PICK, null);
129                intent1.setDataAndType(MediaStore.Images.Media.EXTERNAL_CONTENT_URI,
130                                                                 "image/*");
131                startActivityForResult(intent1, 3);
132                dialog.dismiss();
133            }
134        });
135        //调用照相机
136        tv_select_camera.setOnClickListener(new View.OnClickListener() {
137            @Override
138            public void onClick(View v) {
139                Intent intent2=new Intent(MediaStore.ACTION_IMAGE_CAPTURE);
140                intent2.putExtra(MediaStore.EXTRA_OUTPUT,
141                    Uri.fromFile(new File(Environment.getExternalStorageDirectory(),
142                spUserName+"_head.jpg")));
143                startActivityForResult(intent2, 4);       //采用ForResult打开
144                dialog.dismiss();
145            }
146        });
147        dialog.setView(view);
148        dialog.show();
149    }
150    /**
151     * 设置性别的弹出框
152     */
153    private void sexDialog(String sex){
```

```java
154        int sexFlag=0;
155        if("男".equals(sex)){
156            sexFlag=0;
157        }else if("女".equals(sex)){
158            sexFlag=1;
159        }
160        final String items[]={"男","女"};
161        AlertDialog.Builder builder=new AlertDialog.Builder(this);        //先得到构造器
162        builder.setTitle("性别");        //设置标题
163        builder.setSingleChoiceItems(items,sexFlag,new DialogInterface
164                .OnClickListener() {
165            @Override
166            public void onClick(DialogInterface dialog, int which) {
167                dialog.dismiss();
168                //第二个参数 which 是选中的某一项
169                Toast.makeText(UserInfoActivity.this,items[which],
170                        Toast.LENGTH_SHORT).show();
171                setSex(items[which]);
172            }
173        });
174        builder.create().show();
175    }
176    /**
177     * 更新界面中的性别数据
178     */
179    private void setSex(String sex){
180        tv_sex.setText(sex);
181        //更新数据库中的性别字段
182        DBUtils.getInstance(UserInfoActivity.this).updateUserInfo("sex",sex,
183                spUserName);
184    }
185    /**
186     * 获取回传数据时需使用的跳转方法,第一个参数 to 表示需要跳转到的界面,
187     * 第二个参数 requestCode 表示一个请求码,第三个参数 b 表示跳转时传递的数据
188     */
189    public void enterActivityForResult(Class<?>to, int requestCode, Bundle b) {
190        Intent i=new Intent(this, to);
191        i.putExtras(b);
192        startActivityForResult(i, requestCode);
193    }
194    /**
195     * 回传数据
196     */
197    private String new_info;        //最新数据
198    @Override
199    protected void onActivityResult(int requestCode, int resultCode, Intent data)
200    {
201        super.onActivityResult(requestCode, resultCode, data);
202        switch(requestCode) {
203            case CROP_PHOTO1:
```

```
204                 if(resultCode==RESULT_OK) {
205                     cropPhoto(data.getData());              //裁剪图片
206                 }
207                 break;
208             case CROP_PHOTO2:
209                 if(resultCode==RESULT_OK) {
210                     File temp=new File(Environment.getExternalStorageDirectory()
211                             +"/"+spUserName+"_head.jpg");
212                     cropPhoto(Uri.fromFile(temp));           //裁剪图片
213                 }
214                 break;
215             case SAVE_PHOTO:
216                 if(data !=null) {
217                     Bundle extras=data.getExtras();
218                     head=extras.getParcelable("data");
219                     if(head !=null) {
220                         String fileName=setPicToView(head);   //保存在 SD 卡中
221                         //保存头像地址到数据库
222                         DBUtils.getInstance(UserInfoActivity.this).updateUserInfo(
223                                 "head", fileName,spUserName);
224                         iv_photo.setImageBitmap(head);        //用 ImageView 显示出来
225                         //发送广播更新"我"界面中的头像
226                         Intent intent=new Intent(
227                                 UpdateUserInfoReceiver.ACTION.UPDATE_USERINFO);
228                         intent.putExtra(
229                                 UpdateUserInfoReceiver.INTENT_TYPE.TYPE_NAME,
230                                 UpdateUserInfoReceiver.INTENT_TYPE.UPDATE_HEAD);
231                         intent.putExtra("head", fileName);
232                         sendBroadcast(intent);
233                     }
234                 }
235                 break;
236         }
237     }
238     /**
239      * 调用系统的裁剪功能
240      */
241     public void cropPhoto(Uri uri) {
242         Intent intent=new Intent("com.android.camera.action.CROP");
243         intent.setDataAndType(uri, "image/*");
244         intent.putExtra("crop", "true");
245         //aspectX aspectY 是宽和高的比例
246         intent.putExtra("aspectX", 1);
247         intent.putExtra("aspectY", 1);
248         //outputX outputY 是裁剪图片的宽和高
249         intent.putExtra("outputX", 150);
250         intent.putExtra("outputY", 150);
251         intent.putExtra("return-data", true);
252         startActivityForResult(intent, SAVE_PHOTO);
253     }
```

```
254     private String setPicToView(Bitmap mBitmap) {
255         String sdStatus=Environment.getExternalStorageState();
256         if(!sdStatus.equals(Environment.MEDIA_MOUNTED))  {   //检测sd是否可用
257             return "";
258         }
259         FileOutputStream b=null;
260         File file=new File(path);
261         file.mkdirs();                                       //创建文件夹
262         String fileName=path+spUserName+"_head.jpg";         //图片地址和名称作为文件名
263         try {
264             b=new FileOutputStream(fileName);
265             mBitmap.compress(Bitmap.CompressFormat.JPEG, 100, b);    //把数据写入文件
266         } catch(FileNotFoundException e) {
267             e.printStackTrace();
268         } finally {
269             try {
270                 //关闭流
271                 b.flush();
272                 b.close();
273             } catch(IOException e) {
274                 e.printStackTrace();
275             }
276         }
277         return fileName;
278     }
279 }
```

（4）修改 DBUtils.java 文件。由于"个人资料"界面需要获取、保存以及修改个人资料信息，因此需要在 com.itheima.topline.utils 包中找到 DBUtils 类，在该类中添加如下代码：

```
/**
 * 保存个人资料信息
 */
public void saveUserInfo(UserBean bean) {
    ContentValues cv=new ContentValues();
    cv.put("userName", bean.getUserName());
    cv.put("nickName", bean.getNickName());
    cv.put("sex", bean.getSex());
    cv.put("signature", bean.getSignature());
    db.insert(SQLiteHelper.U_USERINFO, null, cv);
}
/**
 * 获取个人资料信息
 */
public UserBean getUserInfo(String userName) {
    String sql="SELECT * FROM "+SQLiteHelper.U_USERINFO+" WHERE userName=?";
    Cursor cursor=db.rawQuery(sql, new String[]{userName});
    UserBean bean=null;
```

```
    while (cursor.moveToNext()) {
      bean=new UserBean();
      bean.setUserName(cursor.getString(cursor.getColumnIndex("userName")));
      bean.setNickName(cursor.getString(cursor.getColumnIndex("nickName")));
      bean.setSex(cursor.getString(cursor.getColumnIndex("sex")));
      bean.setSignature(cursor.getString(cursor.getColumnIndex("signature")));
      bean.setHead(cursor.getString(cursor.getColumnIndex("head")));
    }
    cursor.close();
    return bean;
}
/**
 * 修改个人资料
 */
public void updateUserInfo(String key, String value, String userName) {
    ContentValues cv=new ContentValues();
    cv.put(key, value);
    db.update(SQLiteHelper.U_USERINFO, cv, "userName=?",new String[]{userName});
}
```

（5）修改清单文件。由于"个人资料"界面向右滑动会关闭该界面，因此需要给该界面添加透明主题的样式，在清单文件的 UserInfoActivity 对应的 activity 标签中添加如下代码：

```
android:theme="@style/AppTheme.TransparentActivity"
```

（6）修改 MeFragment.java 文件。当用户处于登录状态时，点击"我"界面中的头像会跳转到"个人资料"界面，因此需要找到文件 10-10 中的 onClick()方法，在该方法的注释"//跳转到'个人资料'界面"下方添加如下代码：

```
Intent userinfo=new Intent(getActivity(), UserInfoActivity.class);
startActivity(userinfo);
```

（7）添加权限。由于在"个人资料"界面修改头像时会访问 SD 卡，因此需要在清单文件中添加 SD 卡的写权限。具体代码如下：

```
<Uses-permission android:name"android.permission.WRITE_EXTERNAL_STORAGE"/>
```

10.6 个人资料修改

任务综述

"个人资料"修改界面主要用于修改用户昵称和签名，由于"修改昵称"界面和"修改签名"界面基本相同，因此可以使用同一个布局文件，根据"个人资料"界面传递过来的参数 flag 判断修改的是哪个属性。

【任务 10-14】 个人资料修改界面

【任务分析】

个人资料修改界面主要用于修改用户的昵称和签名,界面效果如图 10-5 所示。

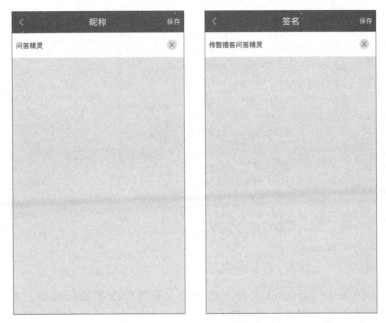

图 10-5　个人资料修改界面

【任务实施】

(1) 创建个人资料修改界面。在 com.itheima.topline.activity 包中创建一个 Empty Activity 类,命名为 ChangeUserInfoActivity 并将布局文件名指定为 activity_change_user_info。在该布局文件中,通过<include>标签将 main_title_bar.xml(标题栏)引入。

(2) 导入界面图片。将个人资料修改界面所需的图片 info_delete.png 导入 drawable-hdpi 文件夹。

(3) 放置界面控件。在布局文件中,放置一个 EditText 控件用于输入文字,一个 ImageView 控件用于显示删除图标,具体代码如文件 10-23 所示。

【文件 10-23】　activity_change_user_info.xml

```
1  <?xml version="1.0" encoding="utf-8"?>
2  <LinearLayout xmlns:android="http://schemas.android.com/apk/res/android"
3      android:layout_width="match_parent"
4      android:layout_height="match_parent"
5      android:background="#eeeeee"
6      android:orientation="vertical">
7      <include layout="@layout/main_title_bar" />
8      <LinearLayout
9          android:layout_width="fill_parent"
```

```
10          android:layout_height="wrap_content"
11          android:gravity="center_vertical"
12          android:orientation="horizontal">
13      <EditText
14          android:id="@+id/et_content"
15          android:layout_width="match_parent"
16          android:layout_height="50dp"
17          android:layout_gravity="center_horizontal"
18          android:background="@android:color/white"
19          android:gravity="center_vertical"
20          android:paddingLeft="10dp"
21          android:singleLine="true"
22          android:textColor="#737373"
23          android:textSize="14sp" />
24      <ImageView
25          android:id="@+id/iv_delete"
26          android:layout_width="27dp"
27          android:layout_height="27dp"
28          android:layout_marginLeft="-40dp"
29          android:src="@drawable/info_delete" />
30      </LinearLayout>
31  </LinearLayout>
```

（4）修改 main_title_bar.xml 文件。由于个人资料修改界面的标题栏右侧有一个"保存"按钮，因此需要在第 6 章中找到文件 6-3，在该文件中添加一个"保存"按钮，具体代码如下所示。

```
<TextView
    android:id="@+id/tv_save"
    android:layout_width="45dp"
    android:layout_height="45dp"
    android:layout_alignParentRight="true"
    android:gravity="center"
    android:text="保存"
    android:textColor="@android:color/white"
    android:textSize="14sp"
    android:visibility="gone" />
```

【任务 10-15】 个人资料修改界面逻辑代码

【任务分析】

在个人资料修改界面中，通过传递过来的标识码判断需要加载修改昵称界面还是修改签名界面。当用户信息修改完成后，点击"保存"按钮将用户信息保存到数据库。需要注意的是，当用户输入昵称或签名时，需要对输入的文字长度进行限制，因此需要给 EditText 控件添加监听事件。

第 10 章 "我"模块(一)

【任务实施】

(1) 获取界面控件。在 ChangeUserInfoActivity 中创建界面控件的初始化方法 init(),用于获取个人资料修改界面所要用到的控件,同时在此方法中还需要设置保存按钮、返回键及删除图标的点击事件。

(2) 监听要修改的文字。由于昵称和签名的长度是有限的,因此需要创建 contentListener() 方法监听输入的文字个数,使昵称不超过 8 个汉字,签名不超过 16 个汉字,具体代码如文件 10-24 所示。

【文件 10-24】 ChangeUserInfoActivity.java

```
1   package com.itheima.topline.activity;
2   public class ChangeUserInfoActivity extends AppCompatActivity {
3       private TextView tv_main_title, tv_save;
4       private RelativeLayout rl_title_bar;
5       private TextView tv_back;
6       private String title, content;
7       private int flag;          //flag 为 1 时表示修改昵称,为 2 时表示修改签名
8       private EditText et_content;
9       private ImageView iv_delete;
10      private SwipeBackLayout layout;
11      @Override
12      protected void onCreate(Bundle savedInstanceState) {
13          super.onCreate(savedInstanceState);
14          layout= (SwipeBackLayout) LayoutInflater.from(this).inflate(
15                  R.layout.base, null);
16          layout.attachToActivity(this);
17          setContentView(R.layout.activity_change_user_info);
18          //设置此界面为竖屏
19          init();
20      }
21      private void init() {
22          //从个人资料界面传递过来的标题和内容
23          title=getIntent().getStringExtra("title");
24          content=getIntent().getStringExtra("content");
25          flag=getIntent().getIntExtra("flag", 0);
26          tv_main_title= (TextView) findViewById(R.id.tv_main_title);
27          tv_main_title.setText(title);
28          rl_title_bar= (RelativeLayout) findViewById(R.id.title_bar);
29          rl_title_bar.setBackgroundColor(getResources().getColor(
30                  R.color.rdTextColorPress));
31          tv_back= (TextView) findViewById(R.id.tv_back);
32          tv_save= (TextView) findViewById(R.id.tv_save);
33          tv_back.setVisibility(View.VISIBLE);
34          tv_save.setVisibility(View.VISIBLE);
35          et_content= (EditText) findViewById(R.id.et_content);
36          iv_delete= (ImageView) findViewById(R.id.iv_delete);
37          if(!TextUtils.isEmpty(content)) {
38              et_content.setText(content);
```

```java
39              et_content.setSelection(content.length());
40          }
41          contentListener();
42          tv_back.setOnClickListener(new View.OnClickListener() {
43              @Override
44              public void onClick(View v) {
45                  ChangeUserInfoActivity.this.finish();
46              }
47          });
48          iv_delete.setOnClickListener(new View.OnClickListener() {
49              @Override
50              public void onClick(View v) {
51                  et_content.setText("");
52              }
53          });
54          tv_save.setOnClickListener(new View.OnClickListener() {
55              @Override
56              public void onClick(View v) {
57                  Intent data=new Intent();
58                  String etContent=et_content.getText().toString().trim();
59                  switch(flag) {
60                      case 1:
61                          if(!TextUtils.isEmpty(etContent)) {
62                              data.putExtra("nickName", etContent);
63                              setResult(RESULT_OK, data);
64                              Toast.makeText(ChangeUserInfoActivity.this, "保存成功",
65                                      Toast.LENGTH_SHORT).show();
66                              ChangeUserInfoActivity.this.finish();
67                          } else {
68                              Toast.makeText(ChangeUserInfoActivity.this,
69                                      "昵称不能为空",Toast.LENGTH_SHORT).show();
70                          }
71                          break;
72                      case 2:
73                          if(!TextUtils.isEmpty(etContent)) {
74                              data.putExtra("signature", etContent);
75                              setResult(RESULT_OK, data);
76                              Toast.makeText(ChangeUserInfoActivity.this, "保存成功",
77                                      Toast.LENGTH_SHORT).show();
78                              ChangeUserInfoActivity.this.finish();
79                          } else {
80                              Toast.makeText(ChangeUserInfoActivity.this,
81                                      "签名不能为空", Toast.LENGTH_SHORT).show();
82                          }
83                          break;
84                  }
85              }
86          });
87      }
```

```java
88      /**
89       * 监听个人资料修改界面输入的文字
90       */
91      private void contentListener() {
92          et_content.addTextChangedListener(new TextWatcher() {
93              @Override
94              public void onTextChanged(CharSequence s, int start, int before,
95                      int count) {
96                  Editable editable=et_content.getText();
97                  int len=editable.length();          //输入文本的长度
98                  if(len>0) {
99                      iv_delete.setVisibility(View.VISIBLE);
100                 } else {
101                     iv_delete.setVisibility(View.GONE);
102                 }
103                 switch(flag) {
104                     case 1:             //昵称
105                         //昵称限制最多8个汉字,超过8个需要截取掉多余的汉字
106                         if(len>8) {
107                             int selEndIndex=Selection.getSelectionEnd(editable);
108                             String str=editable.toString();
109                             //截取新字符串
110                             String newStr=str.substring(0, 8);
111                             et_content.setText(newStr);
112                             editable=et_content.getText();
113                             //新字符串的长度
114                             int newLen=editable.length();
115                             //旧光标位置超过新字符串的长度
116                             if(selEndIndex>newLen) {
117                                 selEndIndex=editable.length();
118                             }
119                             //设置新光标所在的位置
120                             Selection.setSelection(editable, selEndIndex);
121                         }
122                         break;
123                     case 2:             //签名
124                         //签名最多是16个汉字,超过16个需要截取掉多余的汉字
125                         if(len>16) {
126                             int selEndIndex=Selection.getSelectionEnd(editable);
127                             String str=editable.toString();
128                             //截取新字符串
129                             String newStr=str.substring(0, 16);
130                             et_content.setText(newStr);
131                             editable=et_content.getText();
132                             //新字符串的长度
133                             int newLen=editable.length();
134                             //旧光标位置超过新字符串的长度
135                             if(selEndIndex>newLen) {
136                                 selEndIndex=editable.length();
137                             }
```

```
138                            //设置新光标所在的位置
139                            Selection.setSelection(editable, selEndIndex);
140                        }
141                        break;
142                    default:
143                        break;
144                    }
145                }
146                @Override
147                public void beforeTextChanged(CharSequence s, int start, int count,
148                    int after) {
149                }
150                @Override
151                public void afterTextChanged(Editable arg0) {
152                }
153            });
154        }
155    }
```

(3) 修改"个人资料"界面逻辑代码。由于在"个人资料"界面中点击昵称或者签名会跳转到个人资料修改界面，因此需要找到文件 10-22，在"private static final int CROP_PHOTO1=3;"语句下方添加如下代码：

```
private static final int CHANGE_NICKNAME=1;       //修改昵称的自定义常量
private static final int CHANGE_SIGNATURE=2;      //修改签名的自定义常量
```

在该文件的 onClick()方法中的注释"//昵称的点击事件"下方添加如下代码：

```
String name=tv_nickName.getText().toString();     //获取昵称控件上的数据
Bundle bdName=new Bundle();
bdName.putString("content", name);                //传递界面上的昵称数据
bdName.putString("title", "昵称");
bdName.putInt("flag", 1);                         //flag 传递 1 时表示修改昵称
//跳转到个人资料修改界面
enterActivityForResult(ChangeUserInfoActivity.class,CHANGE_NICKNAME, bdName);
```

在该文件的 onClick()方法中的注释"//签名的点击事件"下方添加如下代码：

```
String signature=tv_signature.getText().toString();  //获取签名控件上的数据
Bundle bdSignature=new Bundle();
bdSignature.putString("content", signature);         //传递界面上的签名数据
bdSignature.putString("title", "签名");
bdSignature.putInt("flag", 2);                       //flag 传递 2 时表示修改签名
//跳转到个人资料修改界面
enterActivityForResult(ChangeUserInfoActivity.class,CHANGE_SIGNATURE, bdSignature);
```

在该文件的 onActivityResult()方法中的 switch 语句中添加如下代码：

```
case CHANGE_NICKNAME:           //个人资料修改界面回传过来的昵称数据
    if(data !=null) {
        new_info=data.getStringExtra("nickName");
        if(TextUtils.isEmpty(new_info)) {
```

```
            return;
        }
        tv_nickName.setText(new_info);
        //更新数据库中的昵称字段
        DBUtils.getInstance(UserInfoActivity.this).updateUserInfo(
            "nickName", new_info, spUserName);
    }
    break;
case CHANGE_SIGNATURE:        //个人资料修改界面回传过来的签名数据
    if(data !=null) {
        new_info=data.getStringExtra("signature");
        if(TextUtils.isEmpty(new_info)) {
            return;
        }
        tv_signature.setText(new_info);
        //更新数据库中的签名字段
        DBUtils.getInstance(UserInfoActivity.this).updateUserInfo(
            "signature", new_info, spUserName);
    }
    break;
```

（4）修改清单文件。由于个人资料修改界面向右滑动会关闭该界面，因此需要给该界面添加透明主题的样式，在清单文件的 ChangeUserInfoActivity 对应的 activity 标签中添加如下代码：

```
android:them="@style/AppTheme.TransparentActivity"
```

10.7　本章小结

本章主要讲解了注册、登录、创建 SQLite 数据库、使用数据库存储个人资料以及个人资料的修改。本章涉及数据库的使用，因此建议读者最好温习一下数据库的知识，然后再进行个人资料模块的开发。

【思考题】

1. 如何创建数据库与用户信息表？
2. 如何修改用户头像和性别？

第 11 章
"我"模块(二)

学习目标
- 掌握"日历"界面的开发,使用日历展示当前年份
- 掌握"星座"界面的开发,选择不同的星座展示不同的运势
- 掌握"涂鸦"界面的开发,实现图画的绘制功能
- 掌握"地图"界面的开发,可以定位一个指定地点

"我"模块(二)主要以娱乐趣味为主,该模块包含四个核心功能,分别为日历、星座、涂鸦、地图,让读者在趣味中学习,本章将针对"我"模块(二)进行详细讲解。

11.1 日历

任务综述

思政材料 11

"日历"界面主要是在"我"界面弹出一个对话框,该对话框中显示详细的年、月、日、阴历、阳历、星期以及节假日信息,该界面的日历控件通过调用第三方的 CalendarView 库实现。

【任务 11-1】 "日历"界面

【任务分析】

"日历"界面主要用于方便用户查看日历信息,在该界面中有两个黑色箭头用于切换月份,显示年、月、日的灰色文本、星期信息和对应月份的日期信息,界面效果如图 11-1 所示。

【任务实施】

(1)创建"日历"界面。在 com.itheima.topline.activity 包中创建一个 Empty Activity 类,命名为 CalendarActivity 并将布局文件名指定为 activity_calendar。

(2)导入界面图片。将"日历"界面所需的图片 last.png、next.png 导入 drawable-hdpi 文件夹。

(3)引入 CalendarView 库。该项目中的日历功能是通过引入第三方库 CalendarView 实现的。在 Android Studio 中,选择 File→New→Import Module 选项把日历的框架导入项目,选中项目,右击选择 Open Module Settings 选项,然后选择 Dependencies 选项卡,单击右上角的绿色加号并选择 Module Dependency 选项,把日历框架加入主项目,日历框架详情如图 11-2 所示。

第 11 章 "我"模块(二) 223

图 11-1 "日历"界面

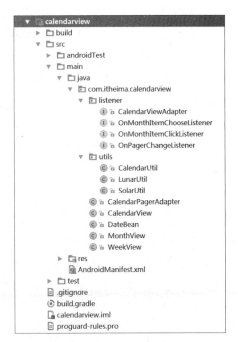

图 11-2 日历框架

(4) 放置界面控件。在布局文件中，放置两个 ImageView 控件，用于显示左右两个黑色箭头；一个 TextView 控件，用于显示年、月、日；一个 WeekView 控件，用于显示星期；一个 CalendarView 控件，用于显示阴历与阳历日期以及一些节假日信息。具体代码如文件 11-1 所示。

【文件 11-1】 activity_calendar.xml

```
1   <?xml version="1.0" encoding="utf-8"?>
2   <LinearLayout xmlns:android="http://schemas.android.com/apk/res/android"
3       xmlns:calendarview="http://schemas.android.com/apk/res-auto"
4       android:layout_width="match_parent"
5       android:layout_height="wrap_content"
6       android:background="#ffffff"
7       android:gravity="center_horizontal"
8       android:orientation="vertical">
9       <RelativeLayout
10          android:layout_width="match_parent"
11          android:layout_height="wrap_content"
12          android:layout_marginTop="15dp">
13          <ImageView
14              android:layout_width="25dp"
15              android:layout_height="25dp"
16              android:layout_marginLeft="50dp"
17              android:background="@drawable/last"
18              android:onClick="lastMonth" />
19          <TextView
20              android:id="@+id/tv_title"
```

```xml
21          android:layout_width="wrap_content"
22          android:layout_height="wrap_content"
23          android:layout_centerInParent="true"
24          android:textColor="@android:color/darker_gray"
25          android:textSize="16sp"
26          android:textStyle="bold" />
27      <ImageView
28          android:layout_width="25dp"
29          android:layout_height="25dp"
30          android:layout_alignParentRight="true"
31          android:layout_marginRight="50dp"
32          android:background="@drawable/next"
33          android:onClick="nextMonth" />
34  </RelativeLayout>
35  <com.itheima.calendarview.WeekView
36      android:layout_width="match_parent"
37      android:layout_height="30dp" />
38  <com.itheima.calendarview.CalendarView
39      android:id="@+id/calendar"
40      android:layout_width="match_parent"
41      android:layout_height="wrap_content"
42      android:padding="4dp"
43      calendarview:date_end="2030.12"
44      calendarview:date_start="1990.1" />
45  </LinearLayout>
```

【任务 11-2】"日历"界面逻辑代码

【任务分析】

"日历"界面主要显示年、月、日、阴历、阳历、星期以及一些节假日信息,当点击"日历"界面中的左右黑色箭头时,会改变显示的月份,点击界面中的日期会显示圆形蓝色背景。

【任务实施】

(1) 获取界面控件。由于该界面是通过调用日历库实现的,所以相对来说比较简单,在 CalendarActivity 的 onCreate()方法中初始化界面控件并设置相应控件的值,具体代码如文件 11-2 所示。

【文件 11-2】 CalendarActivity.java

```java
1   package com.itheima.topline.activity;
2   public class CalendarActivity extends AppCompatActivity {
3       private CalendarView calendarView;
4       private TextView tv_title;
5       @Override
6       protected void onCreate(Bundle savedInstanceState) {
7           super.onCreate(savedInstanceState);
8           setContentView(R.layout.activity_calendar);
```

```
9          tv_title=(TextView) findViewById(R.id.tv_title);
10         calendarView=(CalendarView) findViewById(R.id.calendar);
11         calendarView.init();
12         DateBean d=calendarView.getDateInit();
13         tv_title.setText(d.getSolar()[0]+"年"+d.getSolar()[1]+"月"+
14             d.getSolar()[2]+"日");
15         calendarView.setOnPagerChangeListener(new OnPagerChangeListener() {
16             @Override
17             public void onPagerChanged(int[] date) {
18                 tv_title.setText(date[0]+"年"+date[1]+"月"+date[2]+"日");
19             }
20         });
21         calendarView.setOnItemClickListener(new OnMonthItemClickListener() {
22             @Override
23             public void onMonthItemClick(View view, DateBean date) {
24                 tv_title.setText(date.getSolar()[0]+"年"+date.getSolar()[1]
25                     +"月"+date.getSolar()[2]+"日");
26             }
27         });
28     }
29     public void lastMonth(View view) {
30         calendarView.lastMonth();
31     }
32     public void nextMonth(View view) {
33         calendarView.nextMonth();
34     }
35 }
```

（2）修改 styles.xml 文件。由于"日历"界面是对话框的样式，因此需要在 res/values 文件夹的 styles.xml 文件中添加如下代码：

```
<style name="AppTheme.NoTitle.Dialog" parent="Theme.AppCompat.Dialog">
    <item name="windowActionBar">false</item>
    <item name="windowNoTitle">true</item>
    <item name="android:windowBackground">@android:color/transparent</item>
    <item name="android:windowIsTranslucent">true</item>
</style>
```

（3）修改清单文件。由于"日历"界面是对话框的样式，因此需要给该界面添加对话框的样式，在清单文件的 CalendarActivity 对应的 activity 标签中添加如下代码：

```
android:theme="@style/AppTheme.NoTitle.Dialog"
```

（4）修改"我"界面逻辑代码。由于点击"我"界面中的日历图标时会跳转到"日历"界面，因此需要在第 10 章中找到文件 10-10 中的 onClick() 方法，在该方法中的"case R.id.ll_calendar:"语句下方添加如下代码：

```
Intent calendarIntent=new Intent(getActivity(), CalendarActivity.class);
startActivity(calendarIntent);
```

11.2 星座

任务综述

"星座"界面主要展示十二个星座的详细信息,点击界面右上角的"切换"按钮会切换到不同的星座,切换后的"星座"界面会显示该星座的名称、日期、头像、图标、简介、整体运势、爱情运势、事业学业、财富运势以及健康运势的详细信息。为了界面的美观,在界面右下角会设置不断冒出心形泡泡的效果。

【任务 11-3】 "星座"界面

【任务分析】

"星座"界面主要用于展示被切换的星座的名称、日期、头像、图标、简介、整体运势、爱情运势、事业学业、财富运势、健康运势以及心形泡泡的效果,界面效果如图 11-3 所示。

【任务实施】

(1) 创建"星座"界面。在 com.itheima.topline.activity 包中创建一个 Empty Activity 类,命名为 ConstellationActivity 并将布局文件名指定为 activity_constellation。

(2) 导入界面图片。将"星座"界面所需的图片 constellation_bg.png、whole_icon.png、love_icon.png、career_icon.png、money_icon.png、health_icon.png 导入 drawable-hdpi 文件夹。

(3) 引入 BubbleViews 库。该项目中的心形泡泡效果是通过引入第三方库 BubbleViews 实现的。在 Android Studio 中,选择 File→New→Import Module 选项把心形泡泡的框架导入项目,选中项目,右击选择 Open Module Settings 选项后选择 Dependencies 选项卡,单击右上角的绿色加号并选择 Module Dependency

图 11-3 "星座"界面

选项,把心形泡泡框架加入主项目,心形泡泡框架详情如图 11-4 所示。

(4) 放置界面控件。在布局文件中,放置一个 HeartLayout 自定义控件显示心形泡泡;通过<include>标签将 activity_constellation_content.xml(局部布局)引入,具体代码如文件 11-3 所示。

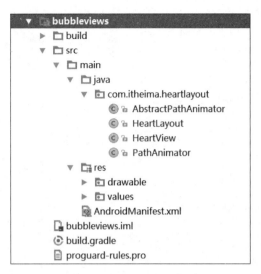

图 11-4　心形泡泡框架

【文件 11-3】　activity_constellation.xml

```
1   <?xml version="1.0" encoding="utf-8"?>
2   <FrameLayout xmlns:android="http://schemas.android.com/apk/res/android"
3       android:layout_width="match_parent"
4       android:layout_height="match_parent"
5       android:background="@android:color/white">
6       <include layout="@layout/activity_constellation_content" />
7       <com.itheima.heartlayout.HeartLayout
8           android:id="@+id/heart_layout"
9           android:layout_width="100dp"
10          android:layout_height="match_parent"
11          android:layout_gravity="bottom|right" />
12  </FrameLayout>
```

（5）放置 activity_constellation_content.xml 文件中的控件。在 res/layout 文件夹中创建一个 activity_constellation_content.xml 文件。在该布局文件中，放置 7 个 ImageView 控件，其中一个 ImageView 控件用于显示星座头像，一个 ImageView 控件用于显示星座图标，剩余 5 个 ImageView 控件分别用于显示整体运势、爱情运势、事业学业、财富运势以及健康运势的图标，放置 17 个 TextView 控件，其中一个 TextView 控件用于显示星座名称，一个 TextView 控件用于显示星座日期，一个 TextView 控件用于显示星座介绍信息，4 个 TextView 控件分别用于显示整体运势、爱情运势、事业学业、财富运势的文本，10 个 TextView 控件分别用于显示整体运势、爱情运势、事业学业、财富运势以及健康运势的文本与详细信息；放置 4 个 RatingBar 控件分别用于显示整体运势、爱情运势、事业学业、财富运势的星级信息。具体代码如文件 11-4 所示。

【文件 11-4】　activity_constellation_content.xml

```
1   <?xml version="1.0" encoding="utf-8"?>
2   <ScrollView xmlns:android="http://schemas.android.com/apk/res/android"
```

```xml
3        android:layout_width="match_parent"
4        android:layout_height="match_parent"
5        android:background="@android:color/white">
6        <LinearLayout
7            android:layout_width="match_parent"
8            android:layout_height="match_parent"
9            android:orientation="vertical">
10           <LinearLayout
11               android:layout_width="match_parent"
12               android:layout_height="180dp"
13               android:layout_alignParentLeft="true"
14               android:layout_alignParentStart="true"
15               android:layout_alignParentTop="true"
16               android:background="@drawable/constellation_bg"
17               android:orientation="vertical">
18               <include layout="@layout/main_title_bar" />
19               <RelativeLayout
20                   android:layout_width="match_parent"
21                   android:layout_height="fill_parent"
22                   android:layout_marginLeft="40dp">
23                   <ImageView
24                       android:id="@+id/iv_head"
25                       android:layout_width="80dp"
26                       android:layout_height="80dp"
27                       android:scaleType="fitXY" />
28                   <LinearLayout
29                       android:layout_width="fill_parent"
30                       android:layout_height="match_parent"
31                       android:layout_marginLeft="15dp"
32                       android:layout_toRightOf="@id/iv_head"
33                       android:orientation="vertical">
34                       <LinearLayout
35                           android:layout_width="fill_parent"
36                           android:layout_height="wrap_content"
37                           android:layout_marginTop="8dp"
38                           android:gravity="center_vertical"
39                           android:orientation="horizontal">
40                           <ImageView
41                               android:id="@+id/iv_icon"
42                               android:layout_width="18dp"
43                               android:layout_height="18dp"
44                               android:scaleType="fitXY" />
45                           <TextView
46                               android:id="@+id/tv_name"
47                               android:layout_width="wrap_content"
48                               android:layout_height="wrap_content"
49                               android:layout_marginLeft="4dp"
50                               android:textColor="@android:color/white"
51                               android:textSize="18sp" />
52                       </LinearLayout>
```

```
53          <TextView
54              android:id="@+id/tv_date"
55              android:layout_width="wrap_content"
56              android:layout_height="wrap_content"
57              android:layout_marginTop="4dp"
58              android:textColor="@android:color/white"
59              android:textSize="18sp" />
60      </LinearLayout>
61      </RelativeLayout>
62  </LinearLayout>
63  <TextView
64      android:id="@+id/tv_info"
65      android:layout_width="fill_parent"
66      android:layout_height="wrap_content"
67      android:layout_marginLeft="10dp"
68      android:layout_marginRight="10dp"
69      android:layout_marginTop="8dp"
70      android:background="@color/constellation_info_bg_color"
71      android:padding="6dp"
72      android:textColor="@color/constellation_info_color"
73      android:textSize="14sp" />
74  <RelativeLayout
75      android:layout_width="wrap_content"
76      android:layout_height="wrap_content"
77      android:padding="10dp">
78      <LinearLayout
79          android:id="@+id/ll_whole"
80          android:layout_width="wrap_content"
81          android:layout_height="wrap_content"
82          android:orientation="horizontal">
83          <TextView
84              android:layout_width="wrap_content"
85              android:layout_height="wrap_content"
86              android:text="整体运势："
87              android:textColor="@color/constellation_info_color2"
88              android:textSize="12sp" />
89          <RatingBar
90              android:id="@+id/rb_whole"
91              style="?android:attr/ratingBarStyleSmall"
92              android:layout_width="wrap_content"
93              android:layout_height="wrap_content"
94              android:layout_gravity="center_vertical"
95              android:progressTint="@color/rating_bar_color"
96              android:rating="2.5" />
97      </LinearLayout>
98      <LinearLayout
99          android:layout_width="wrap_content"
100         android:layout_height="wrap_content"
101         android:layout_alignParentRight="true"
102         android:orientation="horizontal">
```

```xml
103        <TextView
104            android:layout_width="wrap_content"
105            android:layout_height="wrap_content"
106            android:text="爱情运势:"
107            android:textColor="@color/constellation_info_color2"
108            android:textSize="12sp" />
109        <RatingBar
110            android:id="@+id/rb_love"
111            style="?android:attr/ratingBarStyleSmall"
112            android:layout_width="wrap_content"
113            android:layout_height="wrap_content"
114            android:layout_gravity="center_vertical"
115            android:progressTint="@color/rating_bar_color"
116            android:rating="2.5" />
117    </LinearLayout>
118    <LinearLayout
119        android:layout_width="wrap_content"
120        android:layout_height="wrap_content"
121        android:layout_below="@id/ll_whole"
122        android:layout_marginTop="4dp"
123        android:orientation="horizontal">
124        <TextView
125            android:layout_width="wrap_content"
126            android:layout_height="wrap_content"
127            android:text="事业学业:"
128            android:textColor="@color/constellation_info_color2"
129            android:textSize="12sp" />
130        <RatingBar
131            android:id="@+id/rb_career"
132            style="?android:attr/ratingBarStyleSmall"
133            android:layout_width="wrap_content"
134            android:layout_height="wrap_content"
135            android:layout_gravity="center_vertical"
136            android:progressTint="@color/rating_bar_color"
137            android:rating="2.5" />
138    </LinearLayout>
139    <LinearLayout
140        android:layout_width="wrap_content"
141        android:layout_height="wrap_content"
142        android:layout_alignParentRight="true"
143        android:layout_below="@id/ll_whole"
144        android:layout_marginTop="4dp"
145        android:orientation="horizontal">
146        <TextView
147            android:layout_width="wrap_content"
148            android:layout_height="wrap_content"
149            android:text="财富运势:"
150            android:textColor="@color/constellation_info_color2"
151            android:textSize="12sp" />
152        <RatingBar
```

```xml
153                    android:id="@+id/rb_money"
154                    style="?android:attr/ratingBarStyleSmall"
155                    android:layout_width="wrap_content"
156                    android:layout_height="wrap_content"
157                    android:layout_gravity="center_vertical"
158                    android:progressTint="@color/rating_bar_color"
159                    android:rating="2.5" />
160            </LinearLayout>
161        </RelativeLayout>
162        <LinearLayout
163            android:layout_width="fill_parent"
164            android:layout_height="wrap_content"
165            android:layout_marginLeft="10dp"
166            android:orientation="horizontal">
167            <ImageView
168                android:layout_width="20dp"
169                android:layout_height="20dp"
170                android:scaleType="fitXY"
171                android:src="@drawable/whole_icon" />
172            <TextView
173                android:layout_width="wrap_content"
174                android:layout_height="wrap_content"
175                android:layout_marginLeft="4dp"
176                android:text="整体运势"
177                android:textColor="@color/whole_text_color"
178                android:textSize="14sp" />
179        </LinearLayout>
180        <TextView
181            android:id="@+id/tv_whole"
182            android:layout_width="wrap_content"
183            android:layout_height="wrap_content"
184            android:padding="10dp"
185            android:textColor="@color/constellation_info_color2"
186            android:textSize="12sp" />
187        <LinearLayout
188            android:layout_width="fill_parent"
189            android:layout_height="wrap_content"
190            android:layout_marginLeft="10dp"
191            android:orientation="horizontal">
192            <ImageView
193                android:layout_width="20dp"
194                android:layout_height="20dp"
195                android:scaleType="fitXY"
196                android:src="@drawable/love_icon" />
197            <TextView
198                android:layout_width="wrap_content"
199                android:layout_height="wrap_content"
200                android:layout_marginLeft="4dp"
201                android:text="爱情运势"
202                android:textColor="@color/love_text_color"
```

```xml
203                    android:textSize="14sp" />
204            </LinearLayout>
205            <TextView
206                android:id="@+id/tv_love"
207                android:layout_width="wrap_content"
208                android:layout_height="wrap_content"
209                android:padding="10dp"
210                android:textColor="@color/constellation_info_color2"
211                android:textSize="12sp" />
212            <LinearLayout
213                android:layout_width="fill_parent"
214                android:layout_height="wrap_content"
215                android:layout_marginLeft="10dp"
216                android:orientation="horizontal">
217                <ImageView
218                    android:layout_width="20dp"
219                    android:layout_height="20dp"
220                    android:scaleType="fitXY"
221                    android:src="@drawable/career_icon" />
222                <TextView
223                    android:layout_width="wrap_content"
224                    android:layout_height="wrap_content"
225                    android:layout_marginLeft="4dp"
226                    android:text="事业学业"
227                    android:textColor="@color/career_text_color"
228                    android:textSize="14sp" />
229            </LinearLayout>
230            <TextView
231                android:id="@+id/tv_career"
232                android:layout_width="wrap_content"
233                android:layout_height="wrap_content"
234                android:padding="10dp"
235                android:textColor="@color/constellation_info_color2"
236                android:textSize="12sp" />
237            <LinearLayout
238                android:layout_width="fill_parent"
239                android:layout_height="wrap_content"
240                android:layout_marginLeft="10dp"
241                android:orientation="horizontal">
242                <ImageView
243                    android:layout_width="20dp"
244                    android:layout_height="20dp"
245                    android:scaleType="fitXY"
246                    android:src="@drawable/money_icon" />
247                <TextView
248                    android:layout_width="wrap_content"
249                    android:layout_height="wrap_content"
250                    android:layout_marginLeft="4dp"
251                    android:text="财富运势"
252                    android:textColor="@color/money_text_color"
```

```
253                 android:textSize="14sp" />
254             </LinearLayout>
255             <TextView
256                 android:id="@+id/tv_money"
257                 android:layout_width="wrap_content"
258                 android:layout_height="wrap_content"
259                 android:padding="10dp"
260                 android:textColor="@color/constellation_info_color2"
261                 android:textSize="12sp" />
262             <LinearLayout
263                 android:layout_width="fill_parent"
264                 android:layout_height="wrap_content"
265                 android:layout_marginLeft="10dp"
266                 android:orientation="horizontal">
267                 <ImageView
268                     android:layout_width="20dp"
269                     android:layout_height="20dp"
270                     android:scaleType="fitXY"
271                     android:src="@drawable/health_icon" />
272                 <TextView
273                     android:layout_width="wrap_content"
274                     android:layout_height="wrap_content"
275                     android:layout_marginLeft="4dp"
276                     android:text="健康运势"
277                     android:textColor="@color/health_text_color"
278                     android:textSize="14sp" />
279             </LinearLayout>
280             <TextView
281                 android:id="@+id/tv_health"
282                 android:layout_width="wrap_content"
283                 android:layout_height="wrap_content"
284                 android:padding="10dp"
285                 android:textColor="@color/constellation_info_color2"
286                 android:textSize="12sp" />
287     </LinearLayout>
288 </ScrollView>
```

(6) 修改 colors.xml 文件。由于"星座"界面中有几种颜色用于修饰文本或者背景，因此为了便于后续调用，在 res/values 文件夹的 colors.xml 文件中添加如下代码：

```
<color name="constellation_info_color">#7f8080</color>
<color name="constellation_info_bg_color">#f7f2fd</color>
<color name="rating_bar_color">#702ec4</color>
<color name="constellation_info_color2">#454545</color>
<color name="whole_text_color">#7fbee4</color>
<color name="love_text_color">#ff8cb4</color>
<color name="career_text_color">#b29ddd</color>
<color name="money_text_color">#f0c062</color>
<color name="health_text_color">#9cd47a</color>
```

【任务 11-4】 创建 ConstellationBean

【任务分析】

头条项目中的星座属性包括星座 Id、星座选择界面的带色图标、白色星座图标、星座名称、星座日期、星座头像、星座介绍信息、整体运势星级、爱情运势星级、事业学业星级、财富运势星级、整体运势信息、爱情运势信息，为了便于后续对这些属性进行操作，因此创建一个 ConstellationBean 类用于存放这些属性。

【任务实施】

在 com.itheima.topline.bean 包中创建一个 Java 类，命名为 ConstellationBean。在该类中创建星座所需的属性，具体代码如文件 11-5 所示。

【文件 11-5】ConstellationBean.java

```
1   package com.itheima.topline.bean;
2   public class ConstellationBean {
3       private int id;                          //星座 Id
4       private String img;                      //星座选择界面的带色星座图标
5       private String icon;                     //白色星座图标
6       private String name;                     //星座名称
7       private String date;                     //星座日期
8       private String head;                     //星座头像
9       private String info;                     //星座介绍信息
10      private int whole;                       //整体运势星级
11      private int love;                        //爱情运势星级
12      private int career;                      //事业学业星级
13      private int money;                       //财富运势星级
14      private String whole_info;               //整体运势内容
15      private String love_info;                //爱情运势内容
16      private String career_info;              //事业学业内容
17      private String money_info;               //财富运势内容
18      private String health_info;              //健康运势内容
19      public int getId() {
20          return id;
21      }
22      public void setId(int id) {
23          this.id=id;
24      }
25      public String getImg() {
26          return img;
27      }
28      public void setImg(String img) {
29          this.img=img;
30      }
31      public String getName() {
32          return name;
33      }
```

```java
34      public void setName(String name) {
35          this.name=name;
36      }
37      public String getDate() {
38          return date;
39      }
40      public void setDate(String date) {
41          this.date=date;
42      }
43      public String getHead() {
44          return head;
45      }
46      public void setHead(String head) {
47          this.head=head;
48      }
49      public String getInfo() {
50          return info;
51      }
52      public void setInfo(String info) {
53          this.info=info;
54      }
55      public int getWhole() {
56          return whole;
57      }
58      public void setWhole(int whole) {
59          this.whole=whole;
60      }
61      public int getLove() {
62          return love;
63      }
64      public void setLove(int love) {
65          this.love=love;
66      }
67      public int getCareer() {
68          return career;
69      }
70      public void setCareer(int career) {
71          this.career=career;
72      }
73      public int getMoney() {
74          return money;
75      }
76      public void setMoney(int money) {
77          this.money=money;
78      }
79      public String getWhole_info() {
80          return whole_info;
81      }
82      public void setWhole_info(String whole_info) {
83          this.whole_info=whole_info;
84      }
```

```
85      public String getLove_info() {
86          return love_info;
87      }
88      public void setLove_info(String love_info) {
89          this.love_info=love_info;
90      }
91      public String getCareer_info() {
92          return career_info;
93      }
94      public void setCareer_info(String career_info) {
95          this.career_info=career_info;
96      }
97      public String getMoney_info() {
98          return money_info;
99      }
100     public void setMoney_info(String money_info) {
101         this.money_info=money_info;
102     }
103     public String getHealth_info() {
104         return health_info;
105     }
106     public void setHealth_info(String health_info) {
107         this.health_info=health_info;
108     }
109     public String getIcon() {
110         return icon;
111     }
112     public void setIcon(String icon) {
113         this.icon=icon;
114     }
115 }
```

【任务 11-5】 "星座"界面数据

【任务分析】

"星座"界面由"星座图片"与"星座数据"组成,其中图片是通过在 Tomcat 的 ROOT 文件夹中创建一个图片文件夹 constellation 存放的,数据是通过在 ROOT 文件夹中创建一个 constellation_data.json 文件存放的。

【任务实施】

(1) 创建"星座"界面图片存放的文件夹。在 Tomcat 的 ROOT/topline/img 文件夹中创建一个 constellation 文件夹,用于存放"星座"界面的图片。

(2) 在 Tomcat 服务器中创建"星座"界面数据文件。在 Tomcat 的 ROOT/topline 目录中创建一个 constellation_data.json 文件,该文件用于存放"星座"界面需要加载的数据,具体代码如文件 11-6 所示。

【文件 11-6】 constellation_data.json

```
1   [
2     {
3       "id":1,
4       "name":"白羊座",
5       "head":"http://172.16.43.62:8080/topline/img/constellation/baiyang_head_icon.png",
6       "img":"http://172.16.43.62:8080/topline/img/constellation/baiyang_icon.png",
7       "icon":"http://172.16.43.62:8080/topline/img/constellation/baiyang.png",
8       "date":"3.21~4.19",
9       "info":"白羊座的人热情冲动、爱冒险、慷慨、天不怕、地不怕,而且一旦下定决心,不到黄河心不死,
10      排除万难地要达到目的。",
11      "whole":3,
12      "love":3,
13      "career":4,
14      "money":4,
15      "whole_info":"今天对爱情的向往没那么强烈,即使遇到喜欢的人,表现都相比之前要冷静不少,
16      不容易擦出爱情的火花。财运有好转,投资上会有收获,理财规划好,消费上能够精打细算,货比三家。",
17      "love_info":"身边出现不少爱慕者,有伴者,感情生活稳定。",
18      "career_info":"做事效率高,但是有点激进,要注意细节之处。",
19      "money_info":"有得财机会,赚钱轻松,还会得偏财的运气。",
20      "health_info":"要多爱惜身体,不要忽略了健康。"
21    },
22    ......
23    {
24      "id":12,
25      "name":"双鱼座",
26      "head":"http://172.16.43.62:8080/topline/img/constellation/shuangyu_head_icon.png",
27      "img":"http://172.16.43.62:8080/topline/img/constellation/shuangyu_icon.png",
28      "icon":"http://172.16.43.62:8080/topline/img/constellation/shuangyu.png",
29      "date":"2.19~3.20",
30      "info":"双鱼座集合了所有星座的优缺点于一身,同时受水象星座的情绪化影响,使他们原来复杂
31      的性格又添加了更复杂的一笔。双鱼座的人最大的优点是愿意帮助别人,甚至是牺牲自己。",
32      "whole":3,
33      "love":3,
34      "career":3,
35      "money":3,
36      "whole_info":"一个人的时候压抑的情绪特别容易爆发(可能你比较感性吧),可以多接触人群,
37      能够帮助你排解忧郁。工作稳定上升,与同事互动良好,洽谈事务也可在今天进行。对金钱比较迟钝,
38      不要盲目操作。",
39      "love_info":"恋爱中的人别对爱人说谎,心里有事就说出来。",
40      "career_info":"职场很容易妥协、顺从,太软弱易被人欺负。",
41      "money_info":"财运不太好,进账不多,但起码没有金钱麻烦。",
42      "health_info":"身体变得更加健康,小病小痛开始消失。"
43    }
44  ]
```

(3) 解析 JSON 数据。由于从 Tomcat 服务器中获取的 JSON 格式的数据不能直接加载到界面上,因此需要在 com.itheima.topline.utils 包的 JsonParse 类中创建一个

getConstellaList()方法,用于解析"星座"界面获取的 JSON 数据,在 JsonParse 类中需要添加如下代码：

```java
1   public List<ConstellationBean>getConstellaList(String json) {
2       //使用 gson 库解析 JSON 数据
3       Gson gson=new Gson();
4       //创建一个 TypeToken 的匿名子类对象,并调用对象的 getType()方法
5       Type listType=new TypeToken<List<ConstellationBean>>() {
6       }.getType();
7       //把获取到的信息集合存到 constellaList 中
8       List<ConstellationBean> constellaList=gson.fromJson(json, listType);
9       return constellaList;
10  }
```

(4) 创建星座信息表。由于"星座"界面需要根据星座 Id 查询具体星座的详细信息,因此需要把获取的十二星座信息保存到数据库,创建一个 CONSTELLATION 星座信息表,在 com.itheima.topline.sqlite 包的 SQLiteHelper 类中的"public static final String U_USERINFO="userinfo";//用户信息"语句下方添加如下代码：

```java
public static final String CONSTELLATION="constellation";    //十二星座信息
```

在 SQLiteHelper 类中创建一个星座信息表,具体代码如下：

```java
1   /**
2    * 创建十二星座信息表
3    */
4   db.execSQL("CREATE TABLE  IF NOT EXISTS "+CONSTELLATION+"("
5           +"_id INTEGER PRIMARY KEY AUTOINCREMENT, "
6           +"c_id INT, "                   //星座 Id
7           +"name VARCHAR, "               //星座名称
8           +"head VARCHAR, "               //头像
9           +"img VARCHAR,"                 //图标
10          +"icon VARCHAR,"                //白色图标
11          +"date VARCHAR,"                //日期
12          +"info VARCHAR,"                //星座信息
13          +"whole INT,"                   //整体运势
14          +"love INT,"                    //爱情运势
15          +"career INT,"                  //事业学业
16          +"money INT,"                   //财富运势
17          +"whole_info VARCHAR,"          //整体运势信息
18          +"love_info VARCHAR,"           //爱情运势信息
19          +"career_info VARCHAR,"         //事业学业信息
20          +"money_info VARCHAR,"          //财富运势信息
21          +"health_info VARCHAR"          //健康运势信息
22          +")");
```

在 SQLiteHelper 类的 onUpgrade()方法中的"db.execSQL("DROP TABLE IF EXISTS "+U_USERINFO);"语句下方添加如下代码：

```
db.execSQL("DROP TABLE IF EXISTS "+CONSTELLATION);
```

(5)保存星座数据到数据库。由于十二星座信息数据需要保存到数据库,因此在 com. itheima.topline.utils 包的 DBUtils 类中创建一个 saveConstellationInfo()方法保存十二星座的信息数据,具体代码如下所示:

```
1   /**
2    * 保存十二星座信息
3    */
4   public void saveConstellationInfo(List<ConstellationBean>list) {
5       Cursor cursor=db.rawQuery("SELECT * FROM "+SQLiteHelper.CONSTELLATION, null);
6       //添加数据时,如果星座表中有数据,则在添加新数据之前须删除旧数据
7       if(cursor.getCount() !=0)
8       {
9           //删除表中的数据
10          db.execSQL("DELETE FROM "+SQLiteHelper.CONSTELLATION);
11      }
12      for(ConstellationBean bean : list) {
13          ContentValues cv=new ContentValues();
14          cv.put("c_id", bean.getId());
15          cv.put("name", bean.getName());
16          cv.put("head", bean.getHead());
17          cv.put("img", bean.getImg());
18          cv.put("icon", bean.getIcon());
19          cv.put("date", bean.getDate());
20          cv.put("info", bean.getInfo());
21          cv.put("whole", bean.getWhole());
22          cv.put("love", bean.getLove());
23          cv.put("career", bean.getCareer());
24          cv.put("money", bean.getMoney());
25          cv.put("whole_info", bean.getWhole_info());
26          cv.put("love_info", bean.getLove_info());
27          cv.put("career_info", bean.getCareer_info());
28          cv.put("money_info", bean.getMoney_info());
29          cv.put("health_info", bean.getHealth_info());
30          db.insert(SQLiteHelper.CONSTELLATION, null, cv);
31      }
32  }
```

(6)根据星座 Id 从数据库中获取对应星座的信息。由于"星座"界面需要根据星座 Id 查询具体星座的详细信息,因此需要在 com.itheima.topline.utils 包中的 DBUtils 类中创建一个 getConstellationInfo()方法以获取对应星座的信息数据,具体代码如下所示:

```
1   /**
2    * 根据 Id 获取星座信息
3    */
4   public ConstellationBean getConstellationInfo(int c_id) {
5       String sql="SELECT * FROM "+SQLiteHelper.CONSTELLATION+" WHERE c_id=?";
```

```
 6      Cursor cursor=db.rawQuery(sql, new String[]{c_id+""});
 7      ConstellationBean bean=null;
 8      while (cursor.moveToNext()) {
 9          bean=new ConstellationBean();
10          bean.setName(cursor.getString(cursor.getColumnIndex("name")));
11          bean.setHead(cursor.getString(cursor.getColumnIndex("head")));
12          bean.setImg(cursor.getString(cursor.getColumnIndex("img")));
13          bean.setIcon(cursor.getString(cursor.getColumnIndex("icon")));
14          bean.setDate(cursor.getString(cursor.getColumnIndex("date")));
15          bean.setInfo(cursor.getString(cursor.getColumnIndex("info")));
16          bean.setWhole(cursor.getInt(cursor.getColumnIndex("whole")));
17          bean.setLove(cursor.getInt(cursor.getColumnIndex("love")));
18          bean.setCareer(cursor.getInt(cursor.getColumnIndex("career")));
19          bean.setMoney(cursor.getInt(cursor.getColumnIndex("money")));
20          bean.setWhole_info(cursor.getString(cursor.getColumnIndex("whole_info")));
21          bean.setLove_info(cursor.getString(cursor.getColumnIndex("love_info")));
22          bean.setCareer_info(cursor.getString(cursor.getColumnIndex("career_info")));
23          bean.setMoney_info(cursor.getString(cursor.getColumnIndex("money_info")));
24          bean.setHealth_info(cursor.getString(cursor.getColumnIndex("health_info")));
25      }
26      cursor.close();
27      return bean;
28  }
```

（7）修改 Constant.java 文件。在 com.itheima.topline.utils 包中的 Constant 类中添加一个名为 REQUEST_CONSTELLATION_URL 的"星座"界面接口地址，具体代码如下所示：

```
//"星座"界面接口
public static final String REQUEST_CONSTELLATION_URL="/constellation_data.json";
```

【任务 11-6】 "星座"界面逻辑代码

【任务分析】

"星座"界面主要用于展示星座的详细信息，当进入"星座"界面时，首先要从服务器中获取星座信息数据，然后把解析的数据展示到"星座"界面上。当点击右上角的"切换"按钮时，会跳转到"星座选择"界面。

【任务实施】

（1）获取界面控件。在 ConstellationActivity 中创建界面控件的初始化方法 init()，获取"星座"界面所要用到的控件。

（2）获取与设置数据。在 ConstellationActivity 中创建 getData() 与 setData() 方法，分别用于获取服务器中的数据与把获取的数据设置到"星座"界面。

（3）回传数据。在 ConstellationActivity 中重写 onActivityResult() 方法，接收从星座选择界面获取的星座 Id，根据 Id 从数据库获取对应星座信息并展示到界面上，具体代码如文件11-7所示。

【文件11-7】 ConstellationActivity.java

```java
1   package com.itheima.topline.activity;
2   public class ConstellationActivity extends AppCompatActivity {
3       private TextView tv_back, tv_switch;
4       private SwipeBackLayout layout;
5       private ImageView iv_head, iv_icon;
6       private TextView tv_name, tv_date, tv_info;
7       private RatingBar rb_whole, rb_love, rb_career, rb_money;
8       private TextView tv_whole, tv_love, tv_career, tv_money, tv_health;
9       private OkHttpClient okHttpClient;
10      public static final int MSG_CONSTELLATION_OK=1;        //获取星座数据
11      private MHandler mHandler;
12      private Random mRandom;
13      private Timer mTimer;
14      private HeartLayout mHeartLayout;
15      @Override
16      protected void onCreate(Bundle savedInstanceState) {
17          super.onCreate(savedInstanceState);
18          layout= (SwipeBackLayout) LayoutInflater.from(this).inflate(
19                  R.layout.base, null);
20          layout.attachToActivity(this);
21          setContentView(R.layout.activity_constellation);
22          mHandler=new MHandler();
23          okHttpClient=new OkHttpClient();
24          getData();
25          init();
26      }
27      private void init() {
28          mTimer=new Timer();
29          mRandom=new Random();
30          tv_back= (TextView) findViewById(R.id.tv_back);
31          tv_switch= (TextView) findViewById(R.id.tv_save);
32          tv_back.setVisibility(View.VISIBLE);
33          tv_switch.setVisibility(View.VISIBLE);
34          tv_switch.setText("切换");
35          iv_head= (ImageView) findViewById(R.id.iv_head);
36          iv_icon= (ImageView) findViewById(R.id.iv_icon);
37          tv_name= (TextView) findViewById(R.id.tv_name);
38          tv_date= (TextView) findViewById(R.id.tv_date);
39          tv_info= (TextView) findViewById(R.id.tv_info);
40          rb_whole= (RatingBar) findViewById(R.id.rb_whole);
41          rb_love= (RatingBar) findViewById(R.id.rb_love);
42          rb_career= (RatingBar) findViewById(R.id.rb_career);
43          rb_money= (RatingBar) findViewById(R.id.rb_money);
44          tv_whole= (TextView) findViewById(R.id.tv_whole);
45          tv_love= (TextView) findViewById(R.id.tv_love);
46          tv_career= (TextView) findViewById(R.id.tv_career);
47          tv_money= (TextView) findViewById(R.id.tv_money);
```

```
48          tv_health= (TextView) findViewById(R.id.tv_health);
49          tv_back.setOnClickListener(new View.OnClickListener() {
50              @Override
51              public void onClick(View view) {
52                  ConstellationActivity.this.finish();
53              }
54          });
55          tv_switch.setOnClickListener(new View.OnClickListener() {
56              @Override
57              public void onClick(View view) {
58              }
59          });
60          mHeartLayout= (HeartLayout) findViewById(R.id.heart_layout);
61          mTimer.scheduleAtFixedRate(new TimerTask() {
62              @Override
63              public void run() {
64                  mHeartLayout.post(new Runnable() {
65                      @Override
66                      public void run() {
67                          mHeartLayout.addHeart(randomColor());
68                      }
69                  });
70              }
71          }, 500, 200);
72      }
73      private int randomColor() {
74          return Color.rgb(mRandom.nextInt(255), mRandom.nextInt(255),
75          mRandom.nextInt(255));
76      }
77      /**
78       * 事件捕获
79       */
80      class MHandler extends Handler {
81          @Override
82          public void dispatchMessage(Message msg) {
83              super.dispatchMessage(msg);
84              switch(msg.what) {
85                  case MSG_CONSTELLATION_OK:
86                      if(msg.obj !=null) {
87                          String result= (String) msg.obj;
88                          List<ConstellationBean>list=JsonParse.getInstance().
89                          getConstellaList(result);
90                          if(list !=null) {
91                              if(list.size()>0) {
92                                  //保存数据到数据库
93                                  DBUtils.getInstance(ConstellationActivity.this).
94                                                  saveConstellationInfo(list);
95                                  ConstellationBean bean=DBUtils.getInstance(
96                                  ConstellationActivity.this).getConstellationInfo(1);
97                                  setData(bean);
```

```java
98                      }
99                  }
100             }
101             break;
102         }
103     }
104 }
105 private void setData(ConstellationBean bean) {
106     tv_name.setText(bean.getName());
107     Glide
108             .with(ConstellationActivity.this)
109             .load(bean.getHead())
110             .error(R.mipmap.ic_launcher)
111             .into(iv_head);
112     Glide
113             .with(ConstellationActivity.this)
114             .load(bean.getIcon())
115             .error(R.mipmap.ic_launcher)
116             .into(iv_icon);
117     tv_date.setText(bean.getDate());
118     tv_info.setText(bean.getInfo());
119     rb_whole.setRating(bean.getWhole());
120     rb_love.setRating(bean.getLove());
121     rb_career.setRating(bean.getCareer());
122     rb_money.setRating(bean.getMoney());
123     tv_whole.setText(bean.getWhole_info());
124     tv_love.setText(bean.getLove_info());
125     tv_career.setText(bean.getCareer_info());
126     tv_money.setText(bean.getMoney_info());
127     tv_health.setText(bean.getHealth_info());
128 }
129 private void getData() {
130     Request request=new Request.Builder().url(Constant.WEB_SITE+
131     Constant.REQUEST_CONSTELLATION_URL).build();
132     Call call=okHttpClient.newCall(request);
133     //开启异步线程访问网络
134     call.enqueue(new Callback() {
135         @Override
136         public void onResponse(Response response) throws IOException {
137             String res=response.body().string();
138             Message msg=new Message();
139             msg.what=MSG_CONSTELLATION_OK;
140             msg.obj=res;
141             mHandler.sendMessage(msg);
142         }
143         @Override
144         public void onFailure(Request arg0, IOException arg1) {
145         }
146     });
147 }
```

```
148    @Override
149    protected void onDestroy() {
150        super.onDestroy();
151        mTimer.cancel();
152    }
153    @Override
154    protected void onActivityResult(int requestCode, int resultCode, Intent data)
155    {
156        super.onActivityResult(requestCode, resultCode, data);
157        if(data !=null) {
158            int id=data.getIntExtra("id", 0);
159            ConstellationBean bean=DBUtils.getInstance(
160                    ConstellationActivity.this).getConstellationInfo(id);
161            setData(bean);
162        }
163    }
164 }
```

（4）修改清单文件。由于"星座"界面向右滑动会关闭该界面，因此需要给该界面添加透明主题的样式，在清单文件的 ConstellationActivity 对应的 activity 标签中添加如下代码：

```
android:theme="@style/AppTheme.TransparentActivity"
```

（5）修改"我"界面逻辑代码。由于点击"我"界面上的星座图标时会跳转到"星座"界面，因此需要在第 10 章中找到文件 10-10 中的 onClick() 方法，在该方法中的"case R.id.ll_constellation:"语句下方添加如下代码：

```
Intent constellIntent=new Intent(getActivity(), ConstellationActivity.class);
startActivity(constellIntent);
```

11.3 星座选择

任务综述

点击"星座"界面右上角的"切换"按钮会弹出"星座选择"界面，该界面主要用于展示十二星座的图标、名称、阳历日期，点击"星座选择"界面上的任意一个星座，会显示对应星座的详细信息。

【任务 11-7】 "星座选择"界面

【任务分析】

"星座选择"界面主要用于展示十二星座的相关信息，界面效果如图 11-5 所示。

【任务实施】

（1）创建"星座选择"界面。在 com.itheima.topline.activity 包中创建一个 Empty Activity

图 11-5 "星座选择"界面

类,命名为 ChooseConstellationActivity 并将布局文件名指定为 activity_choose_constellation。

(2) 导入界面图片。将"星座选择"界面所需的图片 choose_constella_close_icon.png 导入 drawable-hdpi 文件夹。

(3) 放置界面控件。在布局文件中,放置一个 ImageView 控件用于显示关闭按钮;一个 TextView 控件用于显示"选择星座"文字;一个 View 控件用于显示灰色分隔线;一个 RecyclerView 控件用于显示十二星座的列表信息,具体代码如文件 11-8 所示。

【文件 11-8】 activity_choose_constellation.xml

```
1  <?xml version="1.0" encoding="utf-8"?>
2  <LinearLayout xmlns:android="http://schemas.android.com/apk/res/android"
3      android:layout_width="match_parent"
4      android:layout_height="match_parent"
5      android:background="@android:color/transparent"
6      android:orientation="vertical">
7      <ImageView
8          android:id="@+id/iv_close"
9          android:layout_width="wrap_content"
10         android:layout_height="wrap_content"
11         android:layout_gravity="right"
12         android:src="@drawable/choose_constella_close_icon" />
13     <LinearLayout
14         android:layout_width="240dp"
15         android:layout_height="match_parent"
16         android:background="@android:color/white"
17         android:orientation="vertical">
```

```
18      <TextView
19          android:layout_width="fill_parent"
20          android:layout_height="wrap_content"
21          android:gravity="center"
22          android:padding="8dp"
23          android:text="选择星座"
24          android:textColor="@color/choose_constellation_title_color"
25          android:textSize="14sp" />
26      <View
27          android:layout_width="match_parent"
28          android:layout_height="1dp"
29          android:layout_marginLeft="15dp"
30          android:layout_marginRight="15dp"
31          android:background="@color/divider_line_color" />
32      <android.support.v7.widget.RecyclerView
33          android:id="@+id/rv_list"
34          android:layout_width="fill_parent"
35          android:layout_height="match_parent"
36          android:divider="@null"
37          android:dividerHeight="0dp"
38          android:fadingEdge="none"
39          android:paddingBottom="8dp"
40          android:paddingTop="8dp" />
41  </LinearLayout>
42 </LinearLayout>
```

（4）修改 colors.xml 文件。在 res/values 文件夹的 colors.xml 文件中添加一个名为 choose_constellation_title_color 的颜色值用于设置选择星座文本的颜色，具体代码如下所示：

```
<color name="choose_constellation_title_color">#8bdcea</color>
```

【任务 11-8】 "星座选择"界面 Item

【任务分析】

"星座选择"界面是使用 RecyclerView 控件展示十二星座列表的，因此需要创建一个该列表的 Item 界面。在 Item 界面中需要展示星座图标、名称以及阳历日期，界面效果如图 11-6 所示。

♈ 白羊座 （阳历3.21-4.19）

图 11-6 "星座选择"界面 Item

【任务实施】

（1）创建"星座选择"界面 Item。在 res/layout 文件夹中创建一个布局文件 choose_constella_item.xml。

（2）放置界面控件。在布局文件中，放置一个 ImageView 控件用于显示星座图片；两个 TextView 控件分别用于显示星座名称与星座阳历日期，具体代码如文件 11-9 所示。

【文件 11-9】 choose_constella_item.xml

```xml
1  <?xml version="1.0" encoding="utf-8"?>
2  <LinearLayout xmlns:android="http://schemas.android.com/apk/res/android"
3      android:layout_width="240dp"
4      android:layout_height="25dp"
5      android:layout_marginLeft="30dp"
6      android:background="@android:color/white"
7      android:gravity="center_vertical">
8      <ImageView
9          android:id="@+id/iv_constella_img"
10         android:layout_width="20dp"
11         android:layout_height="20dp"
12         android:scaleType="fitXY" />
13     <TextView
14         android:id="@+id/tv_contella_name"
15         android:layout_width="wrap_content"
16         android:layout_height="wrap_content"
17         android:layout_marginLeft="6dp"
18         android:textColor="@android:color/black"
19         android:textSize="13sp" />
20     <TextView
21         android:id="@+id/tv_date"
22         android:layout_width="wrap_content"
23         android:layout_height="wrap_content"
24         android:layout_marginLeft="6dp"
25         android:textColor="@android:color/black"
26         android:textSize="13sp" />
27 </LinearLayout>
```

【任务 11-9】 "星座选择"界面 Adapter

【任务分析】

"星座选择"列表界面是使用 RecyclerView 控件展示星座信息的,因此需要创建一个数据适配器 ChooseConstellaListAdapter 对 RecyclerView 控件进行数据适配。

【任务实施】

(1) 创建 ChooseConstellaListAdapter 类。在 com.itheima.topline.adapter 包中创建一个 ChooseConstellaListAdapter 类继承 RecyclerView.Adapter<RecyclerView.ViewHolder>类,并重写 onCreateViewHolder()、onBindViewHolder()、getItemCount()方法。

(2) 创建 ViewHolder 类。在 ChooseConstellaListAdapter 类中创建一个 ViewHolder 类用于获取 Item 界面上的控件,具体代码如文件 11-10 所示。

【文件 11-10】 ChooseConstellaListAdapter.java

```java
1   package com.itheima.topline.adapter;
2   public class ChooseConstellaListAdapter extends
3   RecyclerView.Adapter<RecyclerView.ViewHolder> implements View.OnClickListener {
4       private List<ConstellationBean> cbList;
5       private Context context;
6       private OnItemClickListener mOnItemClickListener=null;
7       public ChooseConstellaListAdapter(Context context) {
8           this.context=context;
9       }
10      public void setOnItemClickListener(OnItemClickListener listener) {
11          this.mOnItemClickListener=listener;
12      }
13      public void setData(List<ConstellationBean> cbList) {
14          this.cbList=cbList;
15          notifyDataSetChanged();
16      }
17      @Override
18      public RecyclerView.ViewHolder onCreateViewHolder(ViewGroup viewGroup,
19      int viewType) {
20          View view=LayoutInflater.from(viewGroup.getContext()).inflate(R.layout.
21          choose_constella_item, viewGroup, false);
22          ViewHolder viewHolder=new ViewHolder(view);
23          view.setOnClickListener(this);
24          return viewHolder;
25      }
26      @Override
27      public void onBindViewHolder(final RecyclerView.ViewHolder holder, int i) {
28          ConstellationBean bean=cbList.get(i);
29          ((ViewHolder) holder).tv_contella_name.setText(bean.getName());
30          ((ViewHolder) holder).tv_date.setText(bean.getDate());
31          Glide
32                  .with(context)
33                  .load(bean.getImg())
34                  .error(R.mipmap.ic_launcher)
35                  .into(((ViewHolder) holder).iv_img);
36          //将 i 保存在 itemView 的 Tag 中,以便点击时进行获取
37          holder.itemView.setTag(i);
38      }
39      @Override
40      public int getItemCount() {
41          return cbList==null ? 0 : cbList.size();
42      }
43      @Override
44      public void onClick(View v) {
45          if(mOnItemClickListener !=null) {
46              //注意这里使用 getTag 方法获取 position
47              mOnItemClickListener.onItemClick(v, (int) v.getTag());
48          }
49      }
```

```
50      public class ViewHolder extends RecyclerView.ViewHolder {
51          public TextView tv_contella_name, tv_date;
52          public ImageView iv_img;
53          public ViewHolder(View itemView) {
54              super(itemView);
55              tv_contella_name=(TextView) itemView.findViewById(
56                                              R.id.tv_contella_name);
57              tv_date=(TextView) itemView.findViewById(R.id.tv_date);
58              iv_img=(ImageView) itemView.findViewById(R.id.iv_constella_img);
59          }
60      }
61      public interface OnItemClickListener {
62          void onItemClick(View view, int position);
63      }
64  }
```

【任务 11-10】 "星座选择"界面数据

【任务分析】

"星座选择"界面的图片存放在 Tomcat 的 ROOT 文件夹中的 constellation 文件夹中，数据是通过在 ROOT 文件夹中创建一个 choose_constellation_list_data.json 文件存放的。

【任务实施】

(1) 存放"星座选择"界面的图片。将"星座选择"界面所需要的图片放入 Tomcat 的 ROOT/topline/img/constellation 文件夹中。

(2) 创建"星座选择"界面数据文件。在 Tomcat 的 ROOT/topline 目录中创建一个 choose_constellation_list_data.json 文件，该文件用于存放"星座选择"界面需要加载的数据，具体代码如文件 11-11 所示。

【文件 11-11】 choose_constellation_list_data.json

```
1   [
2     {
3       "id":1,
4       "name":"白羊座",
5       "img":"http://172.16.43.62:8080/topline/img/constellation/baiyang_icon.png",
6       "date":"(阳历 3.21-4.19)"
7     },
8     …
9     {
10      "id":12,
11      "name":"双鱼座",
12      "img":"http://172.16.43.62:8080/topline/img/constellation/shuangyu_icon.png",
13      "date":"(阳历 2.19-3.20)"
14    }
15  ]
```

注意："星座选择"界面与"星座"界面使用的 JSON 解析方法是相同的。

（3）修改 Constant.java 文件。在 com.itheima.topline.utils 包中的 Constant 类中添加一个名为 REQUEST_CHOOSE_CONSTELLATION_LIST_URL 的"星座选择"界面接口地址，具体代码如下所示：

```
//"星座选择"界面接口
public static final String REQUEST_CHOOSE_CONSTELLATION_LIST_URL=
"/choose_constellation_list_data.json";
```

【任务 11-11】 "星座选择"界面逻辑代码

【任务分析】

"星座选择"界面主要显示星座的图标、名称以及阳历日期。点击每个星座会关闭该界面并显示"星座选择"界面，同时会把选择的星座 Id 回传到"星座"界面。

【任务实施】

（1）获取界面控件。在 ChooseConstellationActivity 中创建界面控件的初始化方法 init()，获取"星座选择"界面所要用到的控件。

（2）获取数据。在 ConstellationActivity 中创建 getData() 方法，用于获取服务器中的十二星座信息数据，具体代码如文件 11-12 所示。

【文件 11-12】 ChooseConstellationActivity.java

```
1   package com.itheima.topline.activity;
2   public class ChooseConstellationActivity extends AppCompatActivity {
3       private ImageView iv_close;
4       private ChooseConstellaListAdapter adapter;
5       private OkHttpClient okHttpClient;
6       public static final int MSG_CHOOSE_CONSTELLATION_OK=1;    //获取星座数据
7       private MHandler mHandler;
8       private RecyclerView recyclerView;
9       private List<ConstellationBean>list;
10      @Override
11      protected void onCreate(Bundle savedInstanceState) {
12          super.onCreate(savedInstanceState);
13          setContentView(R.layout.activity_choose_constellation);
14          mHandler=new MHandler();
15          okHttpClient=new OkHttpClient();
16          getData();
17          init();
18      }
19      private void init() {
20          iv_close=(ImageView) findViewById(R.id.iv_close);
21          recyclerView=(RecyclerView) findViewById(R.id.rv_list);
22          //这里用线性显示,类似于 ListView
23          recyclerView.setLayoutManager(new LinearLayoutManager(this));
```

```
24        adapter=new ChooseConstellaListAdapter(ChooseConstellationActivity.this);
25        recyclerView.setAdapter(adapter);
26        iv_close.setOnClickListener(new View.OnClickListener() {
27            @Override
28            public void onClick(View view) {
29                ChooseConstellationActivity.this.finish();
30            }
31        });
32        adapter.setOnItemClickListener(new ChooseConstellaListAdapter.
33        OnItemClickListener() {
34            @Override
35            public void onItemClick(View view, int position) {
36                Intent intent=new Intent();
37                intent.putExtra("id", list.get(position).getId());
38                setResult(RESULT_OK, intent);
39                ChooseConstellationActivity.this.finish();
40            }
41        });
42    }
43    /**
44     * 事件捕获
45     */
46    class MHandler extends Handler {
47        @Override
48        public void dispatchMessage(Message msg) {
49            super.dispatchMessage(msg);
50            switch(msg.what) {
51                case MSG_CHOOSE_CONSTELLATION_OK:
52                    if(msg.obj !=null) {
53                        String result= (String) msg.obj;
54                        list=JsonParse.getInstance().getConstellaList(result);
55                        if(list !=null) {
56                            if(list.size()>0) {
57                                adapter.setData(list);
58                            }
59                        }
60                    }
61                    break;
62            }
63        }
64    }
65    private void getData() {
66        Request request=new Request.Builder().url(Constant.WEB_SITE+
67        Constant.REQUEST_CHOOSE_CONSTELLATION_LIST_URL).build();
68        Call call=okHttpClient.newCall(request);
69        //开启异步线程访问网络
70        call.enqueue(new Callback() {
71            @Override
72            public void onResponse(Response response) throws IOException {
73                String res=response.body().string();
```

```
74                Message msg=new Message();
75                msg.what=MSG_CHOOSE_CONSTELLATION_OK;
76                msg.obj=res;
77                mHandler.sendMessage(msg);
78            }
79            @Override
80            public void onFailure(Request arg0, IOException arg1) {
81            }
82        });
83    }
84 }
```

（3）修改清单文件。由于"星座选择"界面是对话框的样式，因此需要给该界面添加对话框的样式，在清单文件的 ChooseConstellationActivity 对应的 activity 标签中添加如下代码：

```
android:theme="@style/AppTheme.NoTitle.Dialog"
```

（4）修改"星座"界面逻辑代码。由于点击"星座"界面右上角的"切换"按钮会跳转到"星座选择"界面，因此需要找到文件 11-7 中的 init()方法，在该方法中找到 tv_switch 控件的点击事件，并在该控件的点击事件中添加如下代码：

```
Intent intent=new Intent(ConstellationActivity.this,
                ChooseConstellationActivity.class);
startActivityForResult(intent, 1);
```

11.4　涂鸦

任务综述

"涂鸦"界面样式是一个对话框，界面上主要展示"画板""撤销""清除""保存""画笔"以及"橡皮擦"按钮，点击"画笔"可以选择不同颜色与粗细的画笔。点击"撤销""清除""保存""橡皮擦"时，会产生对应功能的效果。

【任务 11-12】　"涂鸦"界面

【任务分析】

"涂鸦"界面主要展示"画板""撤销""清除""保存""画笔"以及"橡皮擦"按钮，界面效果如图 11-7 所示。

【任务实施】

（1）创建"涂鸦"界面。在 com.itheima.topline.activity 包中创建一个 Empty Activity 类，命名为 ScrawActivity 并将布局文件名指定为 activity_scraw。

（2）导入界面图片。将"涂鸦"界面所需的图片 tuya_selectedtrue.png、tuya_

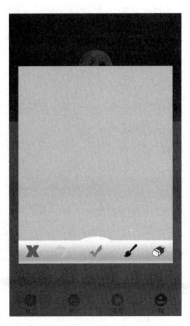

图 11-7 "涂鸦"界面

selectedfalse. png、tuya_toleft. png、tuya_toright. png、tuya_brushsizebg. png、tuya_brushsizeselectedbg. png、tuya_bgbottominit. png、clear_icon. png、undo_icon. png、save_icon. png、pen_icon. png、eraser_icon. png 导入 drawable-hdpi 文件夹。

（3）放置界面控件。该布局文件中的控件相对较多，读者可以从源代码中直接复制，然后梳理布局结构，具体代码如文件 11-13 所示。

【文件 11-13】 activity_scraw. xml

```
1   <?xml version="1.0" encoding="utf-8"?>
2   <RelativeLayout xmlns:android="http://schemas.android.com/apk/res/android"
3       android:layout_width="fill_parent"
4       android:layout_height="400dp"
5       android:orientation="vertical">
6       <FrameLayout
7           android:id="@+id/tuya_layout"
8           android:layout_width="fill_parent"
9           android:layout_height="fill_parent"
10          android:layout_alignParentBottom="true"
11          android:layout_marginBottom="45dp">
12          <ImageView
13              android:id="@+id/imageview_background"
14              android:layout_width="fill_parent"
15              android:layout_height="fill_parent"
16              android:background="#C9DDFE" />
17          <com.itheima.topline.view.ScrawView
18              android:id="@+id/tuyaView"
19              android:layout_width="fill_parent"
20              android:layout_height="fill_parent"
```

```xml
21              android:background="@android:color/transparent" />
22      </FrameLayout>
23      <LinearLayout
24          android:id="@+id/ScrollView01"
25          android:layout_width="fill_parent"
26          android:layout_height="wrap_content"
27          android:layout_alignParentBottom="true"
28          android:layout_marginBottom="46.67dp"
29          android:orientation="vertical"
30          android:visibility="gone">
31          <LinearLayout
32              android:layout_width="fill_parent"
33              android:layout_height="wrap_content"
34              android:orientation="horizontal">
35              <Button
36                  android:id="@+id/colortag"
37                  android:layout_width="fill_parent"
38                  android:layout_height="wrap_content"
39                  android:layout_weight="1"
40                  android:background="@drawable/tuya_selectedtrue"
41                  android:onClick="onClick"
42                  android:text="颜色"
43                  android:textColor="#606060"
44                  android:textSize="16.67dp" />
45              <Button
46                  android:id="@+id/bigtag"
47                  android:layout_width="fill_parent"
48                  android:layout_height="wrap_content"
49                  android:layout_weight="1"
50                  android:background="@drawable/tuya_selectedfalse"
51                  android:onClick="onClick"
52                  android:text="大小"
53                  android:textColor="#606060"
54                  android:textSize="16.67dp" />
55          </LinearLayout>
56          <LinearLayout
57              android:layout_width="fill_parent"
58              android:layout_height="wrap_content"
59              android:background="#c5d0d5"
60              android:orientation="horizontal">
61              <RelativeLayout
62                  android:layout_width="fill_parent"
63                  android:layout_height="wrap_content"
64                  android:layout_marginBottom="12dp"
65                  android:layout_marginTop="7.33dp"
66                  android:orientation="horizontal">
67                  <ImageView
68                      android:id="@+id/imageviewleft"
69                      android:layout_width="wrap_content"
70                      android:layout_height="wrap_content"
```

```xml
71                  android:layout_centerVertical="true"
72                  android:layout_marginLeft="6.67dp"
73                  android:background="@drawable/tuya_toleft" />
74              <ImageView
75                  android:id="@+id/imageviewright"
76                  android:layout_width="wrap_content"
77                  android:layout_height="wrap_content"
78                  android:layout_alignParentRight="true"
79                  android:layout_centerVertical="true"
80                  android:layout_marginRight="6.67dp"
81                  android:background="@drawable/tuya_toright" />
82              <include layout="@layout/tuya_colorlayout" />
83              <!--画笔大小 -->
84              <ScrollView
85                  android:id="@+id/scrollviewbig"
86                  android:layout_width="fill_parent"
87                  android:layout_height="wrap_content"
88                  android:layout_marginLeft="19.33dp"
89                  android:layout_marginRight="19.33dp"
90                  android:layout_marginTop="2.67dip"
91                  android:fadingEdge="none"
92                  android:scrollbars="none"
93                  android:visibility="gone">
94                  <com.itheima.topline.view.MyHorizontalScrollView
95                      android:id="@+id/HorizontalScrollView02"
96                      android:layout_width="fill_parent"
97                      android:layout_height="wrap_content"
98                      android:scrollbars="none">
99                      <LinearLayout
100                         android:id="@+id/LinearLayout02"
101                         android:layout_width="fill_parent"
102                         android:layout_height="wrap_content"
103                         android:orientation="horizontal">
104                         <RelativeLayout
105                             android:layout_width="wrap_content"
106                             android:layout_height="wrap_content"
107                             android:layout_marginLeft="7.33dp"
108                             android:orientation="horizontal">
109                             <Button
110                                 android:id="@+id/sizebutton01"
111                                 android:layout_width="40dp"
112                                 android:layout_height="40dp" />
113                             <LinearLayout
114                                 android:layout_width="40dp"
115                                 android:layout_height="40dp"
116                                 android:orientation="horizontal">
117                                 <ImageView
118                                     android:layout_width="33.33dp"
119                                     android:layout_height="33.33dp"
120                                     android:layout_gravity="center_vertical"
```

```xml
121                    android:layout_marginLeft="3.33dp"
122                    android:background="@drawable/tuya_brushsizebg" />
123                </LinearLayout>
124            </RelativeLayout>
125            <RelativeLayout
126                android:layout_width="wrap_content"
127                android:layout_height="wrap_content"
128                android:layout_marginLeft="7.33dp"
129                android:orientation="horizontal">
130                <Button
131                    android:id="@+id/sizebutton02"
132                    android:layout_width="40dp"
133                    android:layout_height="40dp"
134                    android:background="#aaaa00" />
135                <LinearLayout
136                    android:layout_width="40dp"
137                    android:layout_height="40dp"
138                    android:orientation="horizontal">
139                    <ImageView
140                        android:layout_width="26dp"
141                        android:layout_height="26dp"
142                        android:layout_gravity="center_vertical"
143                        android:layout_marginLeft="6.67dp"
144                        android:background="@drawable/tuya_brushsizebg" />
145                </LinearLayout>
146            </RelativeLayout>
147            <RelativeLayout
148                android:layout_width="wrap_content"
149                android:layout_height="wrap_content"
150                android:layout_marginLeft="7.33dp"
151                android:orientation="horizontal">
152                <Button
153                    android:id="@+id/sizebutton03"
154                    android:layout_width="40dp"
155                    android:layout_height="40dp"
156                    android:background="@null" />
157                <LinearLayout
158                    android:layout_width="40dp"
159                    android:layout_height="40dp"
160                    android:orientation="horizontal">
161                    <ImageView
162                        android:layout_width="20dp"
163                        android:layout_height="20dp"
164                        android:layout_gravity="center_vertical"
165                        android:layout_marginLeft="10dp"
166                        android:background="@drawable/tuya_brushsizebg" />
167                </LinearLayout>
168            </RelativeLayout>
169            <RelativeLayout
170                android:layout_width="wrap_content"
```

```xml
            android:layout_height="wrap_content"
            android:layout_marginLeft="6dp"
            android:orientation="horizontal">
            <Button
                android:id="@+id/sizebutton04"
                android:layout_width="40dp"
                android:layout_height="40dp"
                android:background="@null" />
            <LinearLayout
                android:layout_width="40dp"
                android:layout_height="40dp"
                android:orientation="horizontal">
                <ImageView
                    android:layout_width="13.33dp"
                    android:layout_height="13.33dp"
                    android:layout_gravity="center_vertical"
                    android:layout_marginLeft="13.33dp"
                    android:background="@drawable/tuya_brushsizebg" />
            </LinearLayout>
        </RelativeLayout>
        <RelativeLayout
            android:layout_width="wrap_content"
            android:layout_height="wrap_content"
            android:layout_marginLeft="6dp"
            android:orientation="horizontal">
            <Button
                android:id="@+id/sizebutton05"
                android:layout_width="40dp"
                android:layout_height="40dp"
                android:background="@null" />
            <LinearLayout
                android:layout_width="40dp"
                android:layout_height="40dp"
                android:orientation="horizontal">
                <ImageView
                    android:layout_width="6.67dp"
                    android:layout_height="6.67dp"
                    android:layout_gravity="center_vertical"
                    android:layout_marginLeft="16.67dp"
                    android:background="@drawable/tuya_brushsizebg" />
            </LinearLayout>
        </RelativeLayout>
        <RelativeLayout
            android:layout_width="wrap_content"
            android:layout_height="wrap_content"
            android:layout_marginLeft="5.33dp"
            android:orientation="horizontal">
            <Button
                android:id="@+id/sizebutton06"
                android:layout_width="40dp"
```

```xml
                        android:layout_height="40dp"
                        android:background="@null" />
                    <LinearLayout
                        android:layout_width="40dp"
                        android:layout_height="40dp"
                        android:orientation="horizontal">
                        <ImageView
                            android:layout_width="3.33dp"
                            android:layout_height="3.33dp"
                            android:layout_gravity="center_vertical"
                            android:layout_marginLeft="18.67dp"
                            android:background="@drawable/tuya_brushsizebg" />
                    </LinearLayout>
                </RelativeLayout>
            </LinearLayout>
        </com.itheima.topline.view.MyHorizontalScrollView>
    </ScrollView>
</RelativeLayout>
</LinearLayout>
</LinearLayout>
<LinearLayout
    android:id="@+id/linearlayout"
    android:layout_width="fill_parent"
    android:layout_height="55dp"
    android:layout_alignParentBottom="true"
    android:layout_alignParentLeft="true"
    android:layout_alignParentStart="true"
    android:background="@drawable/tuya_bgbottominit"
    android:orientation="horizontal">
    <LinearLayout
        android:layout_width="fill_parent"
        android:layout_height="wrap_content"
        android:orientation="horizontal">
        <LinearLayout
            android:layout_width="wrap_content"
            android:layout_height="match_parent"
            android:layout_marginTop="15dp"
            android:layout_weight="1"
            android:gravity="center_horizontal|bottom"
            android:orientation="horizontal">
            <LinearLayout
                android:layout_width="fill_parent"
                android:layout_height="wrap_content"
                android:layout_gravity="center"
                android:orientation="vertical">
                <Button
                    android:id="@+id/btn_clear"
                    android:layout_width="30dp"
                    android:layout_height="30dp"
                    android:layout_gravity="center"
```

```
271                android:background="@drawable/clear_icon"
272                android:onClick="onClick" />
273            </LinearLayout>
274        </LinearLayout>
275        <LinearLayout
276            android:layout_width="wrap_content"
277            android:layout_height="match_parent"
278            android:layout_marginTop="15dp"
279            android:layout_weight="1"
280            android:orientation="horizontal">
281            <LinearLayout
282                android:layout_width="fill_parent"
283                android:layout_height="wrap_content"
284                android:layout_gravity="center"
285                android:orientation="vertical">
286                <Button
287                    android:id="@+id/button_undo"
288                    android:layout_width="30dp"
289                    android:layout_height="30dp"
290                    android:layout_gravity="center"
291                    android:background="@drawable/undo_icon"
292                    android:onClick="onClick" />
293            </LinearLayout>
294        </LinearLayout>
295        <LinearLayout
296            android:layout_width="wrap_content"
297            android:layout_height="match_parent"
298            android:layout_marginTop="15dp"
299            android:layout_weight="2"
300            android:orientation="horizontal">
301            <LinearLayout
302                android:layout_width="fill_parent"
303                android:layout_height="wrap_content"
304                android:layout_gravity="center"
305                android:orientation="vertical">
306                <Button
307                    android:id="@+id/btn_save"
308                    android:layout_width="30dp"
309                    android:layout_height="30dp"
310                    android:layout_gravity="center"
311                    android:background="@drawable/save_icon"
312                    android:onClick="onClick" />
313            </LinearLayout>
314        </LinearLayout>
315        <LinearLayout
316            android:layout_width="wrap_content"
317            android:layout_height="match_parent"
318            android:layout_marginTop="15dp"
319            android:layout_weight="1"
320            android:orientation="horizontal">
```

```xml
321         <LinearLayout
322             android:layout_width="fill_parent"
323             android:layout_height="wrap_content"
324             android:layout_gravity="bottom"
325             android:orientation="vertical">
326             <Button
327                 android:id="@+id/button_pen"
328                 android:layout_width="30dp"
329                 android:layout_height="30dp"
330                 android:layout_gravity="center"
331                 android:background="@drawable/pen_icon"
332                 android:onClick="onClick" />
333         </LinearLayout>
334     </LinearLayout>
335     <LinearLayout
336         android:layout_width="wrap_content"
337         android:layout_height="match_parent"
338         android:layout_marginTop="15dp"
339         android:layout_weight="1"
340         android:orientation="horizontal">
341         <LinearLayout
342             android:layout_width="fill_parent"
343             android:layout_height="wrap_content"
344             android:layout_gravity="bottom"
345             android:orientation="vertical">
346             <Button
347                 android:id="@+id/button_eraser"
348                 android:layout_width="30dp"
349                 android:layout_height="30dp"
350                 android:layout_gravity="center_horizontal|bottom"
351                 android:background="@drawable/eraser_icon"
352                 android:onClick="onClick" />
353         </LinearLayout>
354     </LinearLayout>
355     </LinearLayout>
356 </LinearLayout>
357 </RelativeLayout>
```

（4）自定义画布。由于需要设置画布和画笔，并且需要实现撤销、清除、保存、橡皮擦等功能，因此需要在 com.itheima.topline.view 包中创建一个 ScrawView 类继承 View 类，具体代码如文件 11-14 所示。

【文件 11-14】 ScrawView.java

```java
1  package com.itheima.topline.view;
2  public class ScrawView extends View {
3      private Bitmap mBitmap;
4      private Canvas mCanvas;
5      private Path mPath;
```

```java
6      private Paint mBitmapPaint;                                          //画布的画笔
7      private Paint mPaint;                                                //真实的画笔
8      private float mX, mY;                                                //临时点坐标
9      private static final float TOUCH_TOLERANCE=4;
10     //保存 Path 路径的集合,用 List 集合模拟栈,用于后退步骤
11     private static List<DrawPath>savePath;
12     //保存 Path 路径的集合,用 List 集合模拟栈,用于前进步骤
13     private static List<DrawPath>canclePath;
14     private DrawPath dp;                                                 //记录 Path 路径的对象
15     private int screenWidth, screenHeight;                               //屏幕的长和宽
16     private class DrawPath {
17         public Path path;                                                //路径
18         public Paint paint;                                              //画笔
19     }
20     public static int color=Color.parseColor("#fe0000");                 //背景颜色
21     public static int srokeWidth=15;
22     private void init(int w, int h) {
23         screenWidth=w;
24         screenHeight=h;
25         mBitmap=Bitmap.createBitmap(screenWidth, screenHeight,
26              Bitmap.Config.ARGB_8888);
27         mCanvas=new Canvas(mBitmap);                                     //保存每次绘制的图形
28         mBitmapPaint=new Paint(Paint.DITHER_FLAG);
29         initPaint();
30         savePath=new ArrayList<DrawPath>();
31         canclePath=new ArrayList<DrawPath>();
32     }
33     private void initPaint() {
34         mPaint=new Paint();
35         mPaint.setAntiAlias(true);
36         mPaint.setStyle(Paint.Style.STROKE);
37         mPaint.setStrokeJoin(Paint.Join.ROUND);                          //设置外边缘
38         mPaint.setStrokeCap(Paint.Cap.ROUND);                            //形状
39         mPaint.setStrokeWidth(srokeWidth);                               //画笔宽度
40         mPaint.setColor(color);
41     }
42     public ScrawView(Context context, AttributeSet attrs, int defStyle) {
43         super(context, attrs, defStyle);
44         DisplayMetrics dm=new DisplayMetrics();
45         ((Activity) context).getWindowManager().getDefaultDisplay().getMetrics(dm);
46         init(dm.widthPixels, dm.heightPixels);
47     }
48     public ScrawView(Context context, AttributeSet attrs) {
49         super(context, attrs);
50         DisplayMetrics dm=new DisplayMetrics();
51         ((Activity) context).getWindowManager().getDefaultDisplay().getMetrics(dm);
52         init(dm.widthPixels, dm.heightPixels);
53     }
54     @Override
55     public void onDraw(Canvas canvas) {
```

```java
56          //将前面已经画过的内容显示出来
57          canvas.drawBitmap(mBitmap, 0, 0, mBitmapPaint);
58          if(mPath !=null) {
59              //实时显示
60              canvas.drawPath(mPath, mPaint);
61          }
62      }
63      private void touch_start(float x, float y) {
64          mPath.moveTo(x, y);
65          mX=x;
66          mY=y;
67      }
68      private void touch_move(float x, float y) {
69          float dx=Math.abs(x -mX);
70          float dy=Math.abs(mY - y);
71          if(dx>=TOUCH_TOLERANCE || dy>=TOUCH_TOLERANCE) {
72              //从 x1,y1 到 x2,y2 画一条贝塞尔曲线,更平滑(直接用 mPath.lineTo 也可以)
73              mPath.quadTo(mX, mY, (x+mX) / 2, (y+mY) / 2);
74              mX=x;
75              mY=y;
76          }
77      }
78      private void touch_up() {
79          mPath.lineTo(mX, mY);
80          mCanvas.drawPath(mPath, mPaint);
81          //将一条完整的路径保存下来(相当于入栈操作)
82          savePath.add(dp);
83          mPath=null;                         //重新置空
84      }
85      public void clear() {                   //清除画布
86          mBitmap=Bitmap.createBitmap(screenWidth, screenHeight,
87          Bitmap.Config.ARGB_8888);
88          mCanvas.setBitmap(mBitmap);         //重新设置画布,相当于清空画布
89          invalidate();
90      }
91      /**
92       * 撤销的核心思想就是将画布清空,将保存下来的 Path 路径的最后一个移除,重新将路径画在画布上
93       */
94      public int undo() {
95          mBitmap=Bitmap.createBitmap(screenWidth, screenHeight,
96          Bitmap.Config.ARGB_8888);
97          mCanvas.setBitmap(mBitmap);         //重新设置画布,相当于清空画布
98          //清空画布,但如果图片有背景,则使用上面的重新初始化的方法,用该方法会将背景清空
99          if(savePath !=null && savePath.size()>0) {
100             DrawPath dPath=savePath.get(savePath.size() -1);
101             canclePath.add(dPath);
102             //移除最后一个 path,相当于出栈操作
103             savePath.remove(savePath.size() -1);
104             Iterator<DrawPath>iter=savePath.iterator();
105             while (iter.hasNext()) {
```

```java
106            DrawPath drawPath=iter.next();
107            mCanvas.drawPath(drawPath.path, drawPath.paint);
108         }
109         invalidate();         //刷新
110      } else {
111         return -1;
112      }
113      return savePath.size();
114   }
115   public void saveBitmap(Context context) throws Exception {
116      File appDir=new File(Environment.getExternalStorageDirectory(), "tuyaimg");
117      if(!appDir.exists()) {
118         appDir.mkdir();
119      }
120      //产生时间戳,成为文件名
121      String filename=new SimpleDateFormat("yyyyMMddhhmmss",
122         Locale.getDefault()).format(new Date(System.currentTimeMillis()));
123      File file=new File(appDir, filename+".png");
124      file.createNewFile();
125      FileOutputStream fileOutputStream=new FileOutputStream(file);
126      //以100%的品质创建png格式的图片
127      mBitmap.compress(Bitmap.CompressFormat.PNG, 100, fileOutputStream);
128      fileOutputStream.flush();
129      fileOutputStream.close();
130      Toast.makeText(context, "保存成功", Toast.LENGTH_SHORT).show();
131   }
132   @Override
133   public boolean onTouchEvent(MotionEvent event) {
134      float x=event.getX();
135      float y=event.getY();
136      switch(event.getAction()) {
137         case MotionEvent.ACTION_DOWN:
138            initPaint();
139            canclePath=new ArrayList<DrawPath>();         //重置下一步操作
140            mPath=new Path();         //每次按下去重新创建一个Path对象
141            dp=new DrawPath();         //每一次记录的路径对象是不一样的
142            dp.path=mPath;
143            dp.paint=mPaint;
144            touch_start(x, y);
145            invalidate();
146            break;
147         case MotionEvent.ACTION_MOVE:
148            touch_move(x, y);
149            invalidate();
150            break;
151         case MotionEvent.ACTION_UP:
152            touch_up();
153            invalidate();
154            break;
155      }
156      return true;
157   }
158 }
```

（5）自定义水平滑动控件。由于画笔需要水平滑动选择颜色，因此需要在 com.itheima.topline.view 包中创建一个 MyHorizontalScrollView 类继承 HorizontalScrollView 类，具体代码如文件 11-15 所示。

【文件 11-15】 MyHorizontalScrollView.java

```java
1   package com.itheima.topline.view;
2   public class MyHorizontalScrollView extends HorizontalScrollView {
3       public MyHorizontalScrollView(Context context) {
4           super(context);
5       }
6       public MyHorizontalScrollView(Context context, AttributeSet attrs) {
7           super(context, attrs);
8       }
9       public MyHorizontalScrollView(Context context, AttributeSet attrs,
10          int defStyle) {
11          super(context, attrs, defStyle);
12      }
13      @Override
14      protected void onScrollChanged(int l, int t, int oldl, int oldt) {
15          View view= (View) getChildAt(getChildCount() -1);
16          if(view.getLeft() -getScrollX()==0) {       //如果为 0,则证明滑动到最左边
17              onScrollListener.onLeft();
18              Log.d("TAG", "最左边");
19          //如果为 0,则证明滑动到最右边
20          } else if((view.getRight() - (getWidth()+getScrollX()))==0) {
21              onScrollListener.onRight();
22              Log.d("TAG", "最右边");
23          } else {    //说明在中间
24              onScrollListener.onScroll();
25              Log.d("TAG", "中间");
26          }
27          super.onScrollChanged(l, t, oldl, oldt);
28      }
29      private OnScrollListener onScrollListener;
30      public void setOnScrollListener(OnScrollListener onScrollListener) {
31          this.onScrollListener=onScrollListener;
32      }
33      public interface OnScrollListener {
34          void onRight();
35          void onLeft();
36          void onScroll();
37      }
38  }
```

【任务 11-13】 涂鸦颜色选择界面

【任务分析】

在"涂鸦"界面点击画笔会显示画笔颜色与画笔大小的选择界面，点击"颜色"按钮会显示 12 种画笔颜色以供选择，界面效果如图 11-8 所示。

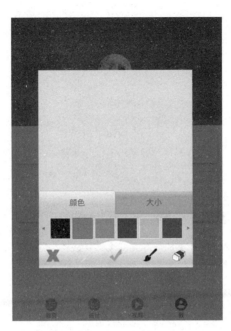

图 11-8 涂鸦颜色选择界面

【任务实施】

（1）创建涂鸦颜色选择界面。在 res/layout 文件夹中创建一个布局文件 tuya_colorlayout.xml。

（2）导入界面图片。将涂鸦颜色选择界面所需的图片 tuya_colorinit.png、tuya_colorselected.png 导入 drawable-hdpi 文件夹。

（3）放置界面控件。由于该布局文件中的控件相对较多，读者可以直接从源代码中复制该布局，梳理布局结构即可，具体代码如文件 11-16 所示。

【文件 11-16】 tuya_colorlayout.xml

```
 1  <?xml version="1.0" encoding="utf-8"?>
 2  <ScrollView xmlns:android="http://schemas.android.com/apk/res/android"
 3      android:id="@+id/scrollviewcolor"
 4      android:layout_width="fill_parent"
 5      android:layout_height="wrap_content"
 6      android:layout_marginLeft="19.33dp"
 7      android:layout_marginRight="19.33dp"
 8      android:fadingEdge="none"
 9      android:scrollbars="none">
10      <com.itheima.topline.view.MyHorizontalScrollView
11          android:id="@+id/HorizontalScrollView01"
12          android:layout_width="fill_parent"
13          android:layout_height="wrap_content"
14          android:scrollbars="none">
15          <LinearLayout
16              android:id="@+id/LinearLayout02"
```

```
17        android:layout_width="fill_parent"
18        android:layout_height="wrap_content"
19        android:orientation="horizontal">
20        1
21        <RelativeLayout
22            android:layout_width="42.67dp"
23            android:layout_height="42.67dp"
24            android:layout_marginLeft="4dp"
25            android:orientation="horizontal">
26            <ImageView
27                android:layout_width="fill_parent"
28                android:layout_height="fill_parent"
29                android:background="@drawable/tuya_colorinit" />
30            <ImageView
31                android:id="@+id/imageview01"
32                android:layout_width="fill_parent"
33                android:layout_height="fill_parent"
34                android:background="@drawable/tuya_colorselected" />
35            <Button
36                android:id="@+id/button01"
37                android:layout_width="fill_parent"
38                android:layout_height="fill_parent"
39                android:layout_margin="1dp"
40                android:background="#140c09" />
41        </RelativeLayout>
42        2
43        <RelativeLayout
44            android:layout_width="42.67dp"
45            android:layout_height="42.67dp"
46            android:layout_marginLeft="4dp"
47            android:orientation="horizontal">
48            <ImageView
49                android:layout_width="fill_parent"
50                android:layout_height="fill_parent"
51                android:background="@drawable/tuya_colorinit" />
52            <ImageView
53                android:id="@+id/imageview02"
54                android:layout_width="fill_parent"
55                android:layout_height="fill_parent"
56                android:background="@drawable/tuya_colorselected" />
57            <Button
58                android:id="@+id/button02"
59                android:layout_width="fill_parent"
60                android:layout_height="fill_parent"
61                android:layout_margin="1dip"
62                android:background="#fe0000" />
63        </RelativeLayout>
64        3
65        <RelativeLayout
66            android:layout_width="42.67dp"
```

```xml
67          android:layout_height="42.67dp"
68          android:layout_marginLeft="4dp"
69          android:orientation="horizontal">
70          <ImageView
71              android:layout_width="fill_parent"
72              android:layout_height="fill_parent"
73              android:background="@drawable/tuya_colorinit" />
74          <ImageView
75              android:id="@+id/imageview03"
76              android:layout_width="fill_parent"
77              android:layout_height="fill_parent"
78              android:background="@drawable/tuya_colorselected" />
79          <Button
80              android:id="@+id/button03"
81              android:layout_width="fill_parent"
82              android:layout_height="fill_parent"
83              android:layout_margin="1dp"
84              android:background="#ff00ea" />
85      </RelativeLayout>
86      4
87      <RelativeLayout
88          android:layout_width="42.67dp"
89          android:layout_height="42.67dp"
90          android:layout_marginLeft="4dp"
91          android:orientation="horizontal">
92          <ImageView
93              android:layout_width="fill_parent"
94              android:layout_height="fill_parent"
95              android:background="@drawable/tuya_colorinit" />
96          <ImageView
97              android:id="@+id/imageview04"
98              android:layout_width="fill_parent"
99              android:layout_height="fill_parent"
100             android:background="@drawable/tuya_colorselected" />
101         <Button
102             android:id="@+id/button04"
103             android:layout_width="fill_parent"
104             android:layout_height="fill_parent"
105             android:layout_margin="1dp"
106             android:background="#011eff" />
107     </RelativeLayout>
108     5
109     <RelativeLayout
110         android:layout_width="42.67dp"
111         android:layout_height="42.67dp"
112         android:layout_marginLeft="4dp"
113         android:orientation="horizontal">
114         <ImageView
115             android:layout_width="fill_parent"
116             android:layout_height="fill_parent"
```

```xml
117                 android:background="@drawable/tuya_colorinit" />
118             <ImageView
119                 android:id="@+id/imageview05"
120                 android:layout_width="fill_parent"
121                 android:layout_height="fill_parent"
122                 android:background="@drawable/tuya_colorselected" />
123             <Button
124                 android:id="@+id/button05"
125                 android:layout_width="fill_parent"
126                 android:layout_height="fill_parent"
127                 android:layout_margin="1dp"
128                 android:background="#00ccff" />
129         </RelativeLayout>
130         6
131         <RelativeLayout
132             android:layout_width="42.67dp"
133             android:layout_height="42.67dp"
134             android:layout_marginLeft="4dp"
135             android:orientation="horizontal">
136             <ImageView
137                 android:layout_width="fill_parent"
138                 android:layout_height="fill_parent"
139                 android:background="@drawable/tuya_colorinit" />
140             <ImageView
141                 android:id="@+id/imageview06"
142                 android:layout_width="fill_parent"
143                 android:layout_height="fill_parent"
144                 android:background="@drawable/tuya_colorselected" />
145             <Button
146                 android:id="@+id/button06"
147                 android:layout_width="fill_parent"
148                 android:layout_height="fill_parent"
149                 android:layout_margin="1dp"
150                 android:background="#00641c" />
151         </RelativeLayout>
152         7
153         <RelativeLayout
154             android:layout_width="42.67dp"
155             android:layout_height="42.67dp"
156             android:layout_marginLeft="4dp"
157             android:orientation="horizontal">
158             <ImageView
159                 android:layout_width="fill_parent"
160                 android:layout_height="fill_parent"
161                 android:background="@drawable/tuya_colorinit" />
162             <ImageView
163                 android:id="@+id/imageview07"
164                 android:layout_width="fill_parent"
165                 android:layout_height="fill_parent"
166                 android:background="@drawable/tuya_colorselected" />
```

```
167        <Button
168            android:id="@+id/button07"
169            android:layout_width="fill_parent"
170            android:layout_height="fill_parent"
171            android:layout_margin="1dp"
172            android:background="#9bff69" />
173    </RelativeLayout>
174    8
175    <RelativeLayout
176        android:layout_width="42.67dp"
177        android:layout_height="42.67dp"
178        android:layout_marginLeft="4dp"
179        android:orientation="horizontal">
180        <ImageView
181            android:layout_width="fill_parent"
182            android:layout_height="fill_parent"
183            android:background="@drawable/tuya_colorinit" />
184        <ImageView
185            android:id="@+id/imageview08"
186            android:layout_width="fill_parent"
187            android:layout_height="fill_parent"
188            android:background="@drawable/tuya_colorselected" />
189        <Button
190            android:id="@+id/button08"
191            android:layout_width="fill_parent"
192            android:layout_height="fill_parent"
193            android:layout_margin="1dp"
194            android:background="#f0ff00" />
195    </RelativeLayout>
196    9
197    <RelativeLayout
198        android:layout_width="42.67dp"
199        android:layout_height="42.67dp"
200        android:layout_marginLeft="4dp"
201        android:orientation="horizontal">
202        <ImageView
203            android:layout_width="fill_parent"
204            android:layout_height="fill_parent"
205            android:background="@drawable/tuya_colorinit" />
206        <ImageView
207            android:id="@+id/imageview09"
208            android:layout_width="fill_parent"
209            android:layout_height="fill_parent"
210            android:background="@drawable/tuya_colorselected" />
211        <Button
212            android:id="@+id/button09"
213            android:layout_width="fill_parent"
214            andrcid:layout_height="fill_parent"
215            android:layout_margin="1dp"
216            android:background="#ff9c00" />
```

```
217        </RelativeLayout>
218        10
219        <RelativeLayout
220            android:layout_width="42.67dp"
221            android:layout_height="42.67dp"
222            android:layout_marginLeft="4dp"
223            android:orientation="horizontal">
224            <ImageView
225                android:layout_width="fill_parent"
226                android:layout_height="fill_parent"
227                android:background="@drawable/tuya_colorinit" />
228            <ImageView
229                android:id="@+id/imageview10"
230                android:layout_width="fill_parent"
231                android:layout_height="fill_parent"
232                android:background="@drawable/tuya_colorselected" />
233            <Button
234                android:id="@+id/button10"
235                android:layout_width="fill_parent"
236                android:layout_height="fill_parent"
237                android:layout_margin="1dp"
238                android:background="#ff5090" />
239        </RelativeLayout>
240        11
241        <RelativeLayout
242            android:layout_width="42.67dp"
243            android:layout_height="42.67dp"
244            android:layout_marginLeft="4dp"
245            android:orientation="horizontal">
246            <ImageView
247                android:layout_width="fill_parent"
248                android:layout_height="fill_parent"
249                android:background="@drawable/tuya_colorinit" />
250            <ImageView
251                android:id="@+id/imageview11"
252                android:layout_width="fill_parent"
253                android:layout_height="fill_parent"
254                android:background="@drawable/tuya_colorselected" />
255            <Button
256                android:id="@+id/button11"
257                android:layout_width="fill_parent"
258                android:layout_height="fill_parent"
259                android:layout_margin="1dp"
260                android:background="#9e9e9e" />
261        </RelativeLayout>
262        12
263        <RelativeLayout
264            android:layout_width="42.67dp"
265            android:layout_height="42.67dp"
266            android:layout_marginLeft="4dp"
```

```
267                 android:orientation="horizontal">
268                     <ImageView
269                         android:layout_width="fill_parent"
270                         android:layout_height="fill_parent"
271                         android:background="@drawable/tuya_colorinit" />
272                     <ImageView
273                         android:id="@+id/imageview12"
274                         android:layout_width="fill_parent"
275                         android:layout_height="fill_parent"
276                         android:background="@drawable/tuya_colorselected" />
277                     <Button
278                         android:id="@+id/button12"
279                         android:layout_width="fill_parent"
280                         android:layout_height="fill_parent"
281                         android:layout_margin="1dp"
282                         android:background="#f5f5f5" />
283                 </RelativeLayout>
284             </LinearLayout>
285         </com.itheima.topline.view.MyHorizontalScrollView>
286 </ScrollView>
```

需要注意的是,由于该布局文件需要显示 12 种颜色,有 12 个类似的相对布局,因此需要在该布局文件中用数字提示这是第几种颜色的布局。

【任务 11-14】 创建 ColorsBean

【任务分析】

在"涂鸦"界面点击"画笔"会显示选择画笔颜色,画笔颜色的属性包括颜色对应的按钮、颜色对应按钮的背景、按钮的 Id 以及颜色的名称。为了便于后续对这些属性进行操作,因此创建一个 ColorsBean 类存放画笔颜色的属性。

【任务实施】

创建 ColorsBean 类。在 com.itheima.topline.bean 包中创建一个 Java 类,命名为 ColorsBean。在该类中创建画笔颜色所需的属性,具体代码如文件 11-17 所示。

【文件 11-17】 ColorsBean.java

```
1   package com.itheima.topline.bean;
2   public class ColorsBean {
3       private Button button;              //颜色对应的按钮
4       private ImageView buttonbg;         //颜色对应的按钮的背景
5       private int tag;                    //按钮的 Id
6       private String name;                //颜色名称
7       public Button getButton() {
8           return button;
9       }
10      public void setButton(Button button) {
```

```
11          this.button=button;
12      }
13      public ImageView getButtonbg() {
14          return buttonbg;
15      }
16      public void setButtonbg(ImageView buttonbg) {
17          this.buttonbg=buttonbg;
18      }
19      public int getTag() {
20          return tag;
21      }
22      public void setTag(int tag) {
23          this.tag=tag;
24      }
25      public String getName() {
26          return name;
27      }
28      public void setName(String name) {
29          this.name=name;
30      }
31  }
```

【任务 11-15】 创建 BigSizeBean

【任务分析】

在"涂鸦"界面点击画笔会显示选择画笔大小,画笔大小的属性包括画笔大小对应的按钮、画笔大小对应按钮的 Id 以及画笔的大小。为了便于后续对这些属性进行操作,因此创建一个 BigSizeBean 类存放画笔大小的属性。

【任务实施】

创建 BigSizeBean 类。在 com.itheima.topline.bean 包中创建一个 Java 类,命名为 BigSizeBean。在该类中创建画笔大小所需的属性,具体代码如文件 11-18 所示。

【文件 11-18】 BigSizeBean.java

```
1   package com.itheima.topline.bean;
2   public class BigSizeBean {
3       private Button button;              //选择画笔大小对应的按钮
4       private int tag;                    //画笔大小对应的按钮的 Id
5       private int name;                   //画笔的大小
6       public Button getButton() {
7           return button;
8       }
9       public void setButton(Button button) {
10          this.button=button;
11      }
12      public int getTag() {
```

```
13              return tag;
14         }
15         public void setTag(int tag) {
16              this.tag=tag;
17         }
18         public int getName() {
19              return name;
20         }
21         public void setName(int name) {
22              this.name=name;
23              int i=Color.RED;
24         }
25    }
```

【任务 11-16】 "涂鸦"界面逻辑代码

【任务分析】

"涂鸦"界面主要展示清除、撤销、保存、画笔、画笔颜色选择、画笔大小选择、橡皮擦等图标,当点击"清除"图标时会清除画板上的涂鸦,点击"撤销""保存""橡皮擦"图标时会分别撤销、保存、擦除画板上的涂鸦,当点击"画笔"图标时,会弹出画笔大小选择以及画笔颜色选择的窗口。

【任务实施】

(1) 获取界面控件。在 ScrawActivity 中创建界面控件的初始化方法 initView(),用于获取涂鸦界面所要用到的控件。

(2) 相关按钮的点击事件。在 ScrawActivity 中实现 OnClickListener 与 OnScrollListener 接口并重写 onClick()方法,在该方法中设置清除、撤销、保存、橡皮擦、画笔、画笔颜色、画笔大小等功能。

(3) 初始化画笔颜色与大小。在 ScrawActivity 中创建 initColourButton()方法用于初始化画笔颜色与画笔大小的数据,具体代码如文件 11-19 所示。

【文件 11-19】 ScrawActivity.java

```
1    package com.itheima.topline.activity;
2    public class ScrawActivity extends AppCompatActivity implements
3    View.OnClickListener, MyHorizontalScrollView.OnScrollListener {
4         private ImageView imageview_background;
5         private ScrawView tuyaView;
6         private FrameLayout tuyaFrameLayout;
7         private static final int UNDO_PATH=1;
8         private static final int USE_ERASER=2;
9         private static final int USE_PAINT=3;
10        private LinearLayout linearlayout;
11        private Button colortag;
12        private Button bigtag;
```

```java
13      private ScrollView scrollviewcolor;
14      private ScrollView scrollviewbig;
15      private MyHorizontalScrollView hscrollViewcolor;
16      private MyHorizontalScrollView hscrollViewsize;
17      private List<ColorsBean> colors=new ArrayList();
18      private ColorsBean color;
19      private List<BigSizeBean> sizes=new ArrayList();
20      private BigSizeBean size;
21      private int index;
22      private int CANCLE_BACKGROUND_IMAGE=0;
23      private final int defaultColor=Color.parseColor("#C9DDFE");
24      @Override
25      protected void onCreate(Bundle savedInstanceState) {
26          super.onCreate(savedInstanceState);
27          setContentView(R.layout.activity_scraw);
28          initView();
29      }
30      private void initView(){
31          tuyaFrameLayout= (FrameLayout) findViewById(R.id.tuya_layout);
32          imageview_background= (ImageView) findViewById(R.id.imageview_background);
33          tuyaView= (ScrawView)findViewById(R.id.tuyaView);
34          colortag= (Button)this.findViewById(R.id.colortag);
35          bigtag= (Button)this.findViewById(R.id.bigtag);
36          scrollviewcolor= (ScrollView)this.findViewById(R.id.scrollviewcolor);
37          scrollviewbig= (ScrollView)this.findViewById(R.id.scrollviewbig);
38          hscrollViewcolor= (MyHorizontalScrollView)this.findViewById(R.id.
39              HorizontalScrollView01);
40          hscrollViewcolor.setOnScrollListener(this);
41          hscrollViewsize= (MyHorizontalScrollView)this.findViewById(R.id.
42              HorizontalScrollView02);
43          hscrollViewsize.setOnScrollListener(this);
44          linearlayout= (LinearLayout)this.findViewById(R.id.ScrollView01);
45          initColorButton();
46      }
47      Handler handler=new Handler() {
48          public void handleMessage(Message msg) {
49              switch(msg.what) {
50                  case UNDO_PATH:
51                      int undo=tuyaView.undo();          //撤销上次的操作
52                      System.out.println("可以撤销:"+undo);
53                      if(undo<0){
54                          CANCLE_BACKGROUND_IMAGE++;
55                          switch(CANCLE_BACKGROUND_IMAGE) {
56                              case 0:
57                                  break;
58                              case 1:
59                                  System.out.println("设置 imageview 为默认");
60                                  imageview_background.setBackgroundColor(defaultColor);
61                                  imageview_background.setImageBitmap(null);
62                                  CANCLE_BACKGROUND_IMAGE=0;
```

```java
63                    break;
64                }
65            }
66            break;
67        case USE_ERASER:
68            if(linearlayout.getVisibility()==View.VISIBLE){
69                linearlayout.setVisibility(View.GONE);
70            }
71            ScrawView.color=Color.parseColor("#C9DDFE");
72            ScrawView.srokeWidth=15;
73            break;
74        case USE_PAINT:
75            if(linearlayout.getVisibility()==View.GONE){
76                linearlayout.setVisibility(View.VISIBLE);
77                ScrawView.srokeWidth=sizes.get(index).getName()+10;
78                for(int i=0;i<colors.size();i++){
79                    if(colors.get(i).getButtonbg().getVisibility()==
80                            View.VISIBLE){
81                        ScrawView.color=Color.parseColor("#"+colors.get(i).
82                                getName());
83                        break;
84                    }
85                }
86            }else{
87                linearlayout.setVisibility(View.GONE);
88            }
89            break;
90        }
91    }
92 };
93 @Override
94 public void onClick(View v) {
95     switch(v.getId()) {
96         case R.id.button_undo:          //撤销
97             Message undo_message=new Message();
98             undo_message.what=UNDO_PATH;
99             handler.sendMessage(undo_message);
100             break;
101        case R.id.button_eraser:        //橡皮擦
102            Message eraser_message=new Message();
103            eraser_message.what=USE_ERASER;
104            handler.sendMessage(eraser_message);
105            break;
106        case R.id.button_pen:           //画笔
107            Message pen_message=new Message();
108            pen_message.what=USE_PAINT;
109            handler.sendMessage(pen_message);
110            break;
111        case R.id.colortag:             //画笔颜色
112            if(linearlayout.getVisibility()==View.VISIBLE){
```

```java
113                 scrollviewcolor.setVisibility(View.VISIBLE);
114                 scrollviewbig.setVisibility(View.GONE);
115                 colortag.setBackgroundResource(R.drawable.tuya_selectedtrue);
116                 bigtag.setBackgroundResource(R.drawable.tuya_selectedfalse);
117             }
118             break;
119         case R.id.bigtag:              //画笔大小
120             if(linearlayout.getVisibility()==View.VISIBLE){
121                 scrollviewcolor.setVisibility(View.GONE);
122                 scrollviewbig.setVisibility(View.VISIBLE);
123                 bigtag.setBackgroundResource(R.drawable.tuya_selectedtrue);
124                 colortag.setBackgroundResource(R.drawable.tuya_selectedfalse);
125             }
126             break;
127         case R.id.btn_clear:           //清除画布
128             tuyaView.clear();
129             break;
130         case R.id.btn_save:            //保存
131             try {
132                 tuyaView.saveBitmap(ScrawActivity.this);         //保存涂鸦信息
133             } catch (Exception e) {
134                 e.printStackTrace();
135             }
136             break;
137     }
138     //颜色
139     int j=-1;
140     for(int i=0;i<colors.size();i++){
141         if(v.getId()==colors.get(i).getTag()){
142             j=i;
143         }
144     }
145     if(j !=-1) {
146         for(int i=0; i<colors.size(); i++) {
147             colors.get(i).getButtonbg().setVisibility(View.INVISIBLE);
148         }
149         colors.get(j).getButtonbg().setVisibility(View.VISIBLE);
150         //改变颜色
151         ScrawView.color=Color.parseColor("#"+colors.get(j).getName());
152         return;
153     }
154     //大小
155     for(int i=0;i<sizes.size();i++){
156         if(v.getId()==sizes.get(i).getTag()){
157             j=i;
158         }
159     }
160     if(j !=-1) {
161         for(int i=0; i<sizes.size(); i++) {
162             sizes.get(i).getButton().setBackgroundResource(0);
163         }
```

```java
164             sizes.get(j).getButton().setBackgroundResource(R.drawable.
165             tuya_brushsizeselectedbg);
166             //改变大小
167             ScrawView.srokeWidth=sizes.get(j).getName()+10;
168             index=j;
169             return;
170         }
171     }
172     @Override
173     public void onRight() {
174     }
175     @Override
176     public void onLeft() {
177     }
178     @Override
179     public void onScroll() {
180     }
181     private void initColorButton() {
182         color=new ColorsBean();
183         color.setButton((Button)this.findViewById(R.id.button01));
184         color.setButtonbg((ImageView)this.findViewById(R.id.imageview01));
185         color.setName("140c09");
186         color.setTag(R.id.button01);
187         colors.add(color);
188         color=new ColorsBean();
189         color.setButton((Button)this.findViewById(R.id.button02));
190         color.setButtonbg((ImageView)this.findViewById(R.id.imageview02));
191         color.setName("fe0000");
192         color.setTag(R.id.button02);
193         colors.add(color);
194         color=new ColorsBean();
195         color.setButton((Button)this.findViewById(R.id.button03));
196         color.setButtonbg((ImageView)this.findViewById(R.id.imageview03));
197         color.setName("ff00ea");
198         color.setTag(R.id.button03);
199         colors.add(color);
200         color=new ColorsBean();
201         color.setButton((Button)this.findViewById(R.id.button04));
202         color.setButtonbg((ImageView)this.findViewById(R.id.imageview04));
203         color.setName("011eff");
204         color.setTag(R.id.button04);
205         colors.add(color);
206         color=new ColorsBean();
207         color.setButton((Button)this.findViewById(R.id.button05));
208         color.setButtonbg((ImageView)this.findViewById(R.id.imageview05));
209         color.setName("00ccff");
210         color.setTag(R.id.button05);
211         colors.add(color);
212         color=new ColorsBean();
213         color.setButton((Button)this.findViewById(R.id.button06));
```

```
214        color.setButtonbg((ImageView)this.findViewById(R.id.imageview06));
215        color.setName("00641c");
216        color.setTag(R.id.button06);
217        colors.add(color);
218        color=new ColorsBean();
219        color.setButton((Button)this.findViewById(R.id.button07));
220        color.setButtonbg((ImageView)this.findViewById(R.id.imageview07));
221        color.setName("9bff69");
222        color.setTag(R.id.button07);
223        colors.add(color);
224        color=new ColorsBean();
225        color.setButton((Button)this.findViewById(R.id.button08));
226        color.setButtonbg((ImageView)this.findViewById(R.id.imageview08));
227        color.setName("f0ff00");
228        color.setTag(R.id.button08);
229        colors.add(color);
230        color=new ColorsBean();
231        color.setButton((Button)this.findViewById(R.id.button09));
232        color.setButtonbg((ImageView)this.findViewById(R.id.imageview09));
233        color.setName("ff9c00");
234        color.setTag(R.id.button09);
235        colors.add(color);
236        color=new ColorsBean();
237        color.setButton((Button)this.findViewById(R.id.button10));
238        color.setButtonbg((ImageView)this.findViewById(R.id.imageview10));
239        color.setName("ff5090");
240        color.setTag(R.id.button10);
241        colors.add(color);
242        color=new ColorsBean();
243        color.setButton((Button)this.findViewById(R.id.button11));
244        color.setButtonbg((ImageView)this.findViewById(R.id.imageview11));
245        color.setName("9e9e9e");
246        color.setTag(R.id.button11);
247        colors.add(color);
248        color=new ColorsBean();
249        color.setButton((Button)this.findViewById(R.id.button12));
250        color.setButtonbg((ImageView)this.findViewById(R.id.imageview12));
251        color.setName("f5f5f5");
252        color.setTag(R.id.button12);
253        colors.add(color);
254        for(int i=0;i<colors.size();i++){
255            colors.get(i).getButton().setOnClickListener(this);
256            colors.get(i).getButtonbg().setVisibility(View.INVISIBLE);
257        }
258        colors.get(1).getButtonbg().setVisibility(View.VISIBLE);
259        size=new BigSizeBean();
260        size.setButton((Button)this.findViewById(R.id.sizebutton01));
261        size.setName(15);
262        size.setTag(R.id.sizebutton01);
263        sizes.add(size);
```

```java
264         size=new BigSizeBean();
265         size.setButton((Button)this.findViewById(R.id.sizebutton02));
266         size.setName(10);
267         size.setTag(R.id.sizebutton02);
268         sizes.add(size);
269         size=new BigSizeBean();
270         size.setButton((Button)this.findViewById(R.id.sizebutton03));
271         size.setName(5);
272         size.setTag(R.id.sizebutton03);
273         sizes.add(size);
274         size=new BigSizeBean();
275         size.setButton((Button)this.findViewById(R.id.sizebutton04));
276         size.setName(0);
277         size.setTag(R.id.sizebutton04);
278         sizes.add(size);
279         size=new BigSizeBean();
280         size.setButton((Button)this.findViewById(R.id.sizebutton05));
281         size.setName(-5);
282         size.setTag(R.id.sizebutton05);
283         sizes.add(size);
284         size=new BigSizeBean();
285         size.setButton((Button)this.findViewById(R.id.sizebutton06));
286         size.setName(-10);
287         size.setTag(R.id.sizebutton06);
288         sizes.add(size);
289         for(int i=0;i<sizes.size();i++){
290             sizes.get(i).getButton().setOnClickListener(this);
291             sizes.get(i).getButton().setBackgroundResource(0);
292         }
293         sizes.get(2).getButton().setBackgroundResource(R.drawable.
294         tuya_brushsizeselectedbg);
295         index=2;
296     }
297     @Override
298     protected void onDestroy() {
299         super.onDestroy();
300         ScrawView.color=Color.parseColor("#fe0000");
301         ScrawView.srokeWidth=15;
302     }
303     @Override
304     public boolean onKeyDown(int keyCode, KeyEvent event) {
305         if(keyCode==KeyEvent.KEYCODE_BACK && event.getRepeatCount()==0) {
306             if(linearlayout.getVisibility()==View.VISIBLE){
307                 linearlayout.setVisibility(View.GONE);
308                 return true;
309             }
310         }
311         return super.onKeyDown(keyCode, event);
312     }
313 }
```

（4）修改清单文件。由于"涂鸦"界面是对话框的样式，因此也需要给该界面添加对话框的样式，在清单文件中，ScrawActivity 对应的 activity 标签中添加如下代码：

```
android:theme="@style/AppTheme.NoTitle.Dialog"
```

（5）修改"我"界面逻辑代码。由于点击"我"界面上的涂鸦图标会跳转到"涂鸦"界面，因此需要在第 10 章中找到文件 10-10 中的 onClick()方法，在该方法中的"case R. id. ll_scraw:"语句下方添加如下代码：

```
Intent scarwIntent=new Intent(getActivity(), ScrawActivity.class);
startActivity(scarwIntent);
```

11.5 地图

任务综述

"地图"界面主要默认设置了一个固定的经纬度，然后在地图上显示该经纬度定位的位置，并调用 Marker 围绕该定位不断地循环移动。

【任务 11-17】 "地图"界面

【任务分析】

"地图"界面主要展示一个地图定位，一个蓝色的 mark 图标会围绕该定位不断地循环移动，界面效果如图 11-9 所示。

图 11-9 "地图"界面

【任务实施】

（1）创建"地图"界面。在 com. itheima. topline. activity 包中创建一个 Empty Activity 类，命名为 MapActivity 并将布局文件名指定为 activity_map。

（2）导入界面图片。将"地图"界面所需的图片 marker. png 导入 drawable-hdpi 文件夹。

（3）添加 Android_3DMap_V2. 4. 0. jar 库。在 Project 选项卡下的 app 中有一个 libs 文件夹，把 Android_3DMap_V2. 4. 0. jar 包复制到 libs 文件夹中，选中 Android_3DMap_V2. 4. 0. jar 包，右击选择 Add As Library 选项，弹出一个对话框，选择把该 jar 包放在 app 的项目中即可。

（4）添加 libamapv304. so 与 libamapv304ex. so 库。由于添加的百度地图的库需要兼容不同的手机，因此需要在 src/main 文件夹中创建一个 jniLibs 文件夹，在该文件夹中分别创建 x86_64、armeabi-v7a、x86、 arm64-v8a、armeabi 文件夹，然后把 libamapv304.so 与 libamapv304ex.so 这两个 so 文件分别放入这 5 个文件夹中，并在 build.gradle 文件中添加如下代码：

```
1  //添加 so 文件需要配置的
2  task nativeLibsToJar(type: Zip, description:
3  "create a jar archive of the native libs") {
4      destinationDir file("$projectDir/libs")
5      baseName "Native_Libs2"
6      extension "jar"
7      from fileTree(dir: "libs", include: "**/*.so")
8      into "lib"
9  }
10 tasks.withType(JavaCompile) {
11     compileTask ->compileTask.dependsOn(nativeLibsToJar)
12 }
```

（5）放置界面控件。在该布局文件中，放置一个 MapView 控件用于显示地图，具体代码如文件 11-20 所示。

【文件 11-20】 activity_map.xml

```
1  <LinearLayout xmlns:android="http://schemas.android.com/apk/res/android"
2      android:layout_width="match_parent"
3      android:layout_height="match_parent"
4      android:orientation="vertical">
5      <include layout="@layout/main_title_bar" />
6      <com.amap.api.maps.MapView
7          android:id="@+id/map"
8          android:layout_width="match_parent"
9          android:layout_height="match_parent"
10         android:layout_weight="1"/>
11 </LinearLayout>
```

【任务 11-18】 "地图"界面逻辑代码

【任务分析】

"地图"界面主要是根据已知的经纬度设置地图的定位，然后设置 marker 的倾斜率，根据倾斜率沿着定位地点慢慢地循环移动。

【任务实施】

（1）获取界面控件。在 MapActivity 中创建界面控件的初始化方法 init()，用于获取地图界面所要用到的控件。

（2）设置 Marker 循环的相关数据。在 MapActivity 中创建 initRoadData()方法，用于设置 Marker 围绕着定位地点循环移动的相关数据。

（3）根据点获取图标转的角度与循环移动。在 MapActivity 中创建 getAngle()方法、moveLooper()方法，分别用于根据点获取图标转的角度与循环移动的逻辑过程，具体代码如文件 11-21 所示。

【文件 11-21】 MapActivity.java

```java
package com.itheima.topline.activity;
public class MapActivity extends AppCompatActivity {
    private MapView mMapView;
    private AMap mAmap;
    private Polyline mVirtureRoad;
    private Marker mMoveMarker;
    //通过设置间隔时间和距离可以控制速度和图标移动的距离
    private static final int TIME_INTERVAL=80;
    private static final double DISTANCE=0.0001;
    private TextView tv_main_title, tv_back;
    private RelativeLayout rl_title_bar;
    private SwipeBackLayout layout;
    @Override
    protected void onCreate(Bundle savedInstanceState) {
        super.onCreate(savedInstanceState);
        layout= (SwipeBackLayout) LayoutInflater.from(this).inflate(
                                                 R.layout.base, null);
        layout.attachToActivity(this);
        setContentView(R.layout.activity_map);
        mMapView= (MapView) findViewById(R.id.map);
        mMapView.onCreate(savedInstanceState);
        mAmap=mMapView.getMap();
        init();
        initRoadData();
        moveLooper();
    }
    private void init() {
        tv_main_title=(TextView) findViewById(R.id.tv_main_title);
        tv_main_title.setText("地图");
        rl_title_bar=(RelativeLayout) findViewById(R.id.title_bar);
        rl_title_bar.setBackgroundColor(getResources().getColor(R.color.
                rdTextColorPress));
        tv_back= (TextView) findViewById(R.id.tv_back);
        tv_back.setVisibility(View.VISIBLE);
        tv_back.setOnClickListener(new View.OnClickListener() {
            @Override
            public void onClick(View view) {
                MapActivity.this.finish();
            }
        });
    }
    private void initRoadData() {
        double centerLatitude=39.916049;
        double centerLontitude=116.399792;
        double deltaAngle=Math.PI / 180 * 5;
        double radius=0.02;
        PolylineOptions polylineOptions=new PolylineOptions();
        for(double i=0; i<Math.PI * 2; i=i+deltaAngle) {
```

```
49              float latitude=(float) (-Math.cos(i) * radius+centerLatitude);
50              float longtitude=(float) (Math.sin(i) * radius+centerLontitude);
51              polylineOptions.add(new LatLng(latitude, longtitude));
52              if(i>Math.PI) {
53                  deltaAngle=Math.PI / 180 * 30;
54              }
55          }
56          float latitude=(float) (-Math.cos(0) * radius+centerLatitude);
57          float longtitude=(float) (Math.sin(0) * radius+centerLontitude);
58          polylineOptions.add(new LatLng(latitude, longtitude));
59          polylineOptions.width(10);
60          polylineOptions.color(Color.RED);
61          mVirtureRoad=mAmap.addPolyline(polylineOptions);
62          MarkerOptions markerOptions=new MarkerOptions();
63          markerOptions.setFlat(true);
64          markerOptions.anchor(0.5f, 0.5f);
65          markerOptions.icon(BitmapDescriptorFactory.fromResource(
66                                          R.drawable.marker));
67          markerOptions.position(polylineOptions.getPoints().get(0));
68          mMoveMarker=mAmap.addMarker(markerOptions);
69          mMoveMarker.setRotateAngle((float) getAngle(0));
70      }
71      /**
72       * 根据点获取图标转的角度
73       */
74      private double getAngle(int startIndex) {
75          if((startIndex+1)>=mVirtureRoad.getPoints().size()) {
76              throw new RuntimeException("index out of bonds");
77          }
78          LatLng startPoint=mVirtureRoad.getPoints().get(startIndex);
79          LatLng endPoint=mVirtureRoad.getPoints().get(startIndex+1);
80          return getAngle(startPoint, endPoint);
81      }
82      /**
83       * 根据两点算取图标转的角度
84       */
85      private double getAngle(LatLng fromPoint, LatLng toPoint) {
86          double slope=getSlope(fromPoint, toPoint);
87          if(slope==Double.MAX_VALUE) {
88              if(toPoint.latitude>fromPoint.latitude) {
89                  return 0;
90              } else {
91                  return 180;
92              }
93          }
94          float deltAngle=0;
95          if((toPoint.latitude - fromPoint.latitude) * slope<0) {
96              deltAngle=180;
97          }
98          double radio=Math.atan(slope);
```

```java
 99            double angle=180 * (radio / Math.PI)+deltAngle - 90;
100            return angle;
101      }
102      /**
103       * 根据点和斜率算取截距
104       */
105      private double getInterception(double slope, LatLng point) {
106            double interception=point.latitude - slope * point.longitude;
107            return interception;
108      }
109      /**
110       * 算斜率
111       */
112      private double getSlope(LatLng fromPoint, LatLng toPoint) {
113            if(toPoint.longitude==fromPoint.longitude) {
114                return Double.MAX_VALUE;
115            }
116            double slope=((toPoint.latitude - fromPoint.latitude) / (
117                          toPoint.longitude - fromPoint.longitude));
118            return slope;
119      }
120      @Override
121      protected void onResume() {
122            super.onResume();
123            mMapView.onResume();
124      }
125      @Override
126      protected void onPause() {
127            super.onPause();
128            mMapView.onPause();
129      }
130      @Override
131      protected void onSaveInstanceState(Bundle outState) {
132            super.onSaveInstanceState(outState);
133            mMapView.onSaveInstanceState(outState);
134      }
135      @Override
136      protected void onDestroy() {
137            super.onDestroy();
138            mMapView.onDestroy();
139      }
140      /**
141       * 计算 x 方向每次移动的距离
142       */
143      private double getXMoveDistance(double slope) {
144            if(slope==Double.MAX_VALUE) {
145                return DISTANCE;
146            }
147            return Math.abs((DISTANCE * slope) / Math.sqrt(1+slope * slope));
148      }
```

```java
149     /**
150      * 循环进行移动逻辑
151      */
152     public void moveLooper() {
153         new Thread() {
154             public void run() {
155                 while (true) {
156                     for(int i=0; i<mVirtureRoad.getPoints().size() -1; i++) {
157                         LatLng startPoint=mVirtureRoad.getPoints().get(i);
158                         LatLng endPoint=mVirtureRoad.getPoints().get(i+1);
159                         mMoveMarker.setPosition(startPoint);
160                         mMoveMarker.setRotateAngle((float) getAngle(startPoint,
161                             endPoint));
162                         double slope=getSlope(startPoint, endPoint);
163                         //是不是正向的标示(向上设为正向)
164                         boolean isReverse=(startPoint.latitude>endPoint.latitude);
165                         double intercept=getInterception(slope, startPoint);
166                         double xMoveDistance=isReverse ? getXMoveDistance(slope)
167                             : -1 * getXMoveDistance(slope);
168                         for(double j=startPoint.latitude;
169                             !((j>endPoint.latitude) ^ isReverse);
170                             j=j -xMoveDistance) {
171                             LatLng latLng=null;
172                             if(slope !=Double.MAX_VALUE) {
173                                 latLng=new LatLng(j, (j -intercept) / slope);
174                             } else {
175                                 latLng=new LatLng(j, startPoint.longitude);
176                             }
177                             mMoveMarker.setPosition(latLng);
178                             try {
179                                 Thread.sleep(TIME_INTERVAL);
180                             } catch(InterruptedException e) {
181                                 e.printStackTrace();
182                             }
183                         }
184                     }
185                 }
186             }
187         }.start();
188     }
189 }
```

（4）修改清单文件。由于"地图"界面向右滑动会关闭该界面，因此需要给该界面添加透明主题的样式，在清单文件的 MapActivity 对应的 activity 标签中添加如下代码：

```
android:theme="@style/AppTheme.TransparentActivity"
```

根据百度地图的官方文档提示，在清单文件的＜application＞节点中需要添加如下代码：

```
<meta-data
    android:name="com.amap.api.v2.apikey"
    android:value="86e5eedc821be0edd51fa307a5da15d3"/>
```

需要注意的是,该地图功能无须申请百度地图密钥,但建议用户申请一个。

(5)修改"我"界面逻辑代码。由于点击"我"界面上的地图图标时会跳转到"地图"界面,因此需要在第 10 章中找到文件 10-10 中的 onClick()方法,在该方法中的"case R.id.ll_map:"语句下方添加如下代码:

```
Intent mapIntent=new Intent(getActivity(), MapActivity.class);
startActivity(mapIntent);
```

11.6 本章小结

本章主要讲解了"我"模块中的日历、星座、涂鸦以及地图等功能。读者通过对本章的学习可以掌握日历、星座、涂鸦和地图界面的搭建过程以及界面开发。

【思考题】

1. 如何实现日历功能?
2. 如何实现地图功能?

第 12 章 设置模块

学习目标

- 掌握"收藏"界面的开发，能够实现新闻收藏功能
- 掌握"设置"界面的开发，能够设置用户的基本信息
- 掌握"修改密码"界面的开发，能够实现密码的修改功能
- 掌握"设置密保"界面的开发，能够实现密保的设置功能

设置模块主要用于实现新闻收藏功能以及用户信息的设置，其中用户信息设置包含三个部分，分别为修改密码、设置密保、退出登录，本章将针对设置模块进行详细讲解。

12.1 收藏

任务综述

思政材料 12

"收藏"界面主要展示用户登录成功后收藏的一些新闻信息，并且侧滑每条信息会出现红色"删除"按钮，点击"删除"按钮会删除对该新闻信息的收藏，点击每条新闻信息会跳转到新闻详情界面。

【任务 12-1】 "收藏"界面

【任务分析】

"收藏"界面主要以列表形式展示用户登录成功后收藏的一些新闻信息，当侧滑每条新闻信息时会出现红色"删除"按钮，界面效果如图 12-1 所示。

【任务实施】

(1) 创建"收藏"界面。在 com. itheima. topline. activity 包中创建一个 Empty Activity 类，命名为 CollectionActivity 并将布局文件名指定为 activity_collection。

(2) 放置界面控件。在该布局文件中，放置一个 RecyclerView 控件用于显示收藏列表；一个 TextView 控件用于显示没有数据时的提示文本，具体代码如文件 12-1 所示。

【文件 12-1】 activity_collection.xml

```
1  <?xml version="1.0" encoding="utf-8"?>
2  <RelativeLayout xmlns:android="http://schemas.android.com/apk/res/android"
3      android:id="@+id/activity_collection"
```

```xml
4      android:layout_width="match_parent"
5      android:layout_height="match_parent">
6      <LinearLayout
7          android:layout_width="match_parent"
8          android:layout_height="match_parent"
9          android:orientation="vertical">
10         <include layout="@layout/main_title_bar" />
11         <android.support.v7.widget.RecyclerView
12             android:id="@+id/rv_recyclerView"
13             android:layout_width="match_parent"
14             android:layout_height="wrap_content"
15             android:background="#EEEEEE" />
16     </LinearLayout>
17     <TextView
18         android:id="@+id/tv_none"
19         android:layout_width="match_parent"
20         android:layout_height="match_parent"
21         android:gravity="center"
22         android:text="暂无收藏信息"
23         android:textColor="@android:color/darker_gray"
24         android:textSize="16sp"
25         android:visibility="gone" />
26 </RelativeLayout>
```

图 12-1 "收藏"界面

【任务 12-2】 "收藏"界面 Item

【任务分析】

由于"收藏"界面用到了 RecyclerView 控件,因此需要为该控件创建一个 Item 界面,界面效果如图 12-2 所示。

图 12-2 "收藏"界面 Item

【任务实施】

(1) 创建"收藏"界面 Item。在 res/layout 文件夹中创建一个布局文件 collection_item.xml。

(2) 放置界面控件。在布局文件中,放置 3 个 TextView 控件,其中一个 TextView 控件用于显示新闻名称;一个 TextView 控件用于显示新闻类型;一个 TextView 控件用于显示删除文字;放置一个 ImageView 控件用于显示新闻图片,具体代码如文件 12-2 所示。

【文件 12-2】 collection_item.xml

```
1   <?xml version="1.0" encoding="utf-8"?>
2   <com.itheima.topline.view.SlidingButtonView
3       xmlns:android="http://schemas.android.com/apk/res/android"
4       android:layout_width="match_parent"
5       android:layout_height="100dp"
6       android:layout_marginBottom="1dp"
7       android:background="@android:color/white">
8       <RelativeLayout
9           android:layout_width="match_parent"
10          android:layout_height="match_parent">
11          <TextView
12              android:id="@+id/tv_delete"
13              android:layout_width="80dp"
14              android:layout_height="match_parent"
15              android:layout_toRightOf="@+id/layout_content"
16              android:background="@drawable/collection_red_del_selector"
17              android:gravity="center"
18              android:text="删 除"
19              android:textColor="#DDFFFFFF" />
20          <RelativeLayout
21              android:id="@+id/layout_content"
22              android:layout_width="match_parent"
23              android:layout_height="match_parent"
24              android:background="@drawable/collection_item_selector"
25              android:padding="8dp">
```

```
26        <ImageView
27            android:id="@+id/iv_img"
28            android:layout_width="100dp"
29            android:layout_height="80dp"
30            android:layout_alignParentLeft="true"
31            android:layout_centerVertical="true"
32            android:scaleType="fitXY" />
33        <LinearLayout
34            android:layout_width="wrap_content"
35            android:layout_height="wrap_content"
36            android:layout_centerVertical="true"
37            android:layout_marginLeft="10dp"
38            android:layout_toRightOf="@id/iv_img"
39            android:orientation="vertical">
40            <TextView
41                android:id="@+id/tv_name"
42                android:layout_width="fill_parent"
43                android:layout_height="wrap_content"
44                android:textColor="@android:color/black"
45                android:textSize="14sp" />
46            <TextView
47                android:id="@+id/tv_newsType_name"
48                android:layout_width="fill_parent"
49                android:layout_height="wrap_content"
50                android:layout_marginTop="8dp"
51                android:textSize="12sp" />
52        </LinearLayout>
53    </RelativeLayout>
54 </RelativeLayout>
55 </com.itheima.topline.view.SlidingButtonView>
```

（3）自定义侧滑控件。由于向左滑动"收藏"界面的 Item 时会出现红色"删除"按钮，因此需要自定义一个侧滑控件 SlidingButtonView 用于"收藏"界面的 Item 布局中，在 com.itheima.topline.view 文件夹中创建一个 SlidingButtonView 类继承 HorizontalScrollView 类，具体代码如文件 12-3 所示。

【文件 12-3】 SlidingButtonView.java

```
1  package com.itheima.topline.view;
2  public class SlidingButtonView extends HorizontalScrollView {
3      private TextView mTextView_Delete;
4      private int mScrollWidth;
5      private IonSlidingButtonListener mIonSlidingButtonListener;
6      private Boolean isOpen=false;
7      private Boolean once=false;
8      public SlidingButtonView(Context context) {
9          this(context, null);
10     }
```

```java
11    public SlidingButtonView(Context context, AttributeSet attrs) {
12        this(context, attrs, 0);
13    }
14    public SlidingButtonView(Context context, AttributeSet attrs,
15    int defStyleAttr) {
16        super(context, attrs, defStyleAttr);
17        this.setOverScrollMode(OVER_SCROLL_NEVER);
18    }
19    @Override
20    protected void onMeasure(int widthMeasureSpec, int heightMeasureSpec) {
21        super.onMeasure(widthMeasureSpec, heightMeasureSpec);
22        if(!once) {
23            mTextView_Delete= (TextView) findViewById(R.id.tv_delete);
24            once=true;
25        }
26    }
27    @Override
28    protected void onLayout(boolean changed, int l, int t, int r, int b) {
29        super.onLayout(changed, l, t, r, b);
30        if(changed) {
31            this.scrollTo(0, 0);
32            //获取水平滚动条可以滑动的范围,即右侧按钮的宽度
33            mScrollWidth=mTextView_Delete.getWidth();
34        }
35    }
36    @Override
37    public boolean onTouchEvent(MotionEvent ev) {
38        int action=ev.getAction();
39        switch(action) {
40            case MotionEvent.ACTION_DOWN:
41            case MotionEvent.ACTION_MOVE:
42                mIonSlidingButtonListener.onDownOrMove(this);
43                break;
44            case MotionEvent.ACTION_UP:
45            case MotionEvent.ACTION_CANCEL:
46                changeScrollx();
47                return true;
48            default:
49                break;
50        }
51        return super.onTouchEvent(ev);
52    }
53    @Override
54    protected void onScrollChanged(int l, int t, int oldl, int oldt) {
55        super.onScrollChanged(l, t, oldl, oldt);
56        mTextView_Delete.setTranslationX(l -mScrollWidth);
57    }
```

```java
58      /**
59       * 按滚动条被拖动的距离判断关闭或打开菜单
60       */
61      public void changeScrollx() {
62          if(getScrollX()>=(mScrollWidth / 2)) {
63              this.smoothScrollTo(mScrollWidth, 0);
64              isOpen=true;
65              mIonSlidingButtonListener.onMenuIsOpen(this);
66          } else {
67              this.smoothScrollTo(0, 0);
68              isOpen=false;
69          }
70      }
71      /**
72       * 打开菜单
73       */
74      public void openMenu() {
75          if(isOpen) {
76              return;
77          }
78          this.smoothScrollTo(mScrollWidth, 0);
79          isOpen=true;
80          mIonSlidingButtonListener.onMenuIsOpen(this);
81      }
82      /**
83       * 关闭菜单
84       */
85      public void closeMenu() {
86          if(!isOpen) {
87              return;
88          }
89          this.smoothScrollTo(0, 0);
90          isOpen=false;
91      }
92      public void setSlidingButtonListener(IonSlidingButtonListener listener)
93      {
94          mIonSlidingButtonListener=listener;
95      }
95      public interface IonSlidingButtonListener {
96          void onMenuIsOpen(View view);
97          void onDownOrMove(SlidingButtonView slidingButtonView);
98      }
100 }
```

（4）创建"删除"按钮的背景选择器。在 res/drawable 文件夹中创建"删除"按钮的背景选择器 collection_red_del_selector.xml，当"删除"按钮被按下时显示深红色背景，当"删除"

按钮弹起时显示浅红色背景，具体代码如文件 12-4 所示。

【文件 12-4】 collection_red_del_selector.xml

```xml
1  <?xml version="1.0" encoding="utf-8"?>
2  <selector xmlns:android="http://schemas.android.com/apk/res/android">
3      <item android:state_pressed="false">
4          <shape>
5              <solid android:color="#FF0000" />
6          </shape>
7      </item>
8      <item android:state_pressed="true">
9          <shape>
10             <solid android:color="#EE1111" />
11         </shape>
12     </item>
13 </selector>
```

（5）创建 Item 的背景选择器。在 res/drawable 文件夹中创建 Item 的背景选择器 collection_btn_black_background.xml。当 Item 被按下时显示灰色背景，当 Item 弹起时显示白色背景，具体代码如文件 12-5 所示。

【文件 12-5】 collection_btn_black_background.xml

```xml
1  <?xml version="1.0" encoding="utf-8"?>
2  <selector xmlns:android="http://schemas.android.com/apk/res/android">
3      <item android:state_pressed="true">
4          <layer-list>
5              <item>
6                  <shape>
7                      <solid android:color="@android:color/white" />
8                  </shape>
9              </item>
10             <item>
11                 <shape>
12                     <solid android:color="#22000000" />
13                 </shape>
14             </item>
15         </layer-list>
16     </item>
17     <item>
18         <shape>
19             <solid android:color="@android:color/white" />
20         </shape>
21     </item>
22 </selector>
```

【任务 12-3】 "收藏"界面 Adapter

【任务分析】

"收藏"界面是通过 RecyclerView 控件展示收藏信息的，因此需要创建一个数据适配器 CollectionAdapter 对 RecyclerView 控件进行数据适配。

【任务实施】

(1) 创建 CollectionAdapter 类。在 com.itheima.topline.adapter 包中，创建一个 CollectionAdapter 类继承 RecyclerView.Adapter<CollectionAdapter.MyViewHolder>类 并实现 SlidingButtonView.IonSlidingButtonListener 接口。

(2) 创建 ViewHolder 类。在 CollectionAdapter 类中创建一个 MyViewHolder 类获取 Item 界面上的控件。

(3) 创建删除收藏条目的方法。在 CollectionAdapter 类中创建一个 removeData()方法用于删除对应的收藏信息，具体代码如文件 12-6 所示。

【文件 12-6】 CollectionAdapter.java

```
1    package com.itheima.topline.adapter;
2    public class CollectionAdapter extends RecyclerView.Adapter<CollectionAdapter.
3    MyViewHolder> implements SlidingButtonView.IonSlidingButtonListener {
4        private Context mContext;
5        private IonSlidingViewClickListener mIDeleteBtnClickListener;
6        private List<NewsBean> newsList=new ArrayList<>();
7        private SlidingButtonView mMenu=null;
8        public CollectionAdapter(Context context) {
9            mContext=context;
10           mIDeleteBtnClickListener= (IonSlidingViewClickListener) context;
11       }
12       public void setData(List<NewsBean> newsList) {
13           this.newsList=newsList;
14           notifyDataSetChanged();
15       }
16       @Override
17       public int getItemCount() {
18           return newsList.size();
19       }
20       @Override
21       public void onBindViewHolder(final MyViewHolder holder, int position) {
22           NewsBean bean=newsList.get(position);
23           holder.tv_name.setText(bean.getNewsName());
24           holder.tv_newsTypeName.setText(bean.getNewsTypeName());
25           Glide
26                   .with(mContext)
27                   .load(bean.getImg1())
28                   .error(R.mipmap.ic_launcher)
29                   .into((holder).iv_img);
```

```java
30              //设置内容布局的宽为屏幕宽度
31              holder.layout_content.getLayoutParams().width=UtilsHelper.
32              getScreenWidth(mContext);
33              holder.layout_content.setOnClickListener(new View.OnClickListener()
34              {
35                  @Override
36                  public void onClick(View v) {
37                      //判断是否有删除菜单打开
38                      if(menuIsOpen()) {
39                          closeMenu();        //关闭菜单
40                      } else {
41                          int n=holder.getLayoutPosition();
42                          mIDeleteBtnClickListener.onItemClick(v, n);
43                      }
44                  }
45              });
46              holder.btn_Delete.setOnClickListener(new View.OnClickListener() {
47                  @Override
48                  public void onClick(View v) {
49                      int n=holder.getLayoutPosition();
50                      mIDeleteBtnClickListener.onDeleteBtnCilck(v, n);
51                  }
52              });
53          }
54          @Override
55          public MyViewHolder onCreateViewHolder(ViewGroup arg0, int arg1) {
56              View view=LayoutInflater.from(mContext).inflate(
57                                  R.layout.collection_item, arg0, false);
58              MyViewHolder holder=new MyViewHolder(view);
59              return holder;
60          }
61          class MyViewHolder extends RecyclerView.ViewHolder {
62              private ImageView iv_img;
63              public TextView btn_Delete, tv_name, tv_newsTypeName;
64              public ViewGroup layout_content;
65              public MyViewHolder(View itemView) {
66                  super(itemView);
67                  btn_Delete=(TextView) itemView.findViewById(R.id.tv_delete);
68                  layout_content=(ViewGroup) itemView.findViewById(
69                                      R.id.layout_content);
70                  iv_img=(ImageView) itemView.findViewById(R.id.iv_img);
71                  tv_name=(TextView) itemView.findViewById(R.id.tv_name);
72                  tv_newsTypeName=(TextView) itemView.findViewById(
73                                      R.id.tv_newsType_name);
74                  ((SlidingButtonView) itemView).setSlidingButtonListener(
75                      CollectionAdapter.this);
76              }
77          }
78          public void removeData(int position, TextView tv_none, String userName)
79          {
```

```java
80          NewsBean bean=newsList.get(position);
81          //从收藏新闻的数据库中也要删除此数据
82          DBUtils.getInstance(mContext).delCollectionNewsInfo(bean.getId(),
83                                                  bean.getType(),userName);
84          newsList.remove(position);
85          notifyItemRemoved(position);
86          if(newsList.size()==0)
87              tv_none.setVisibility(View.VISIBLE);
88      }
89      /**
90       * 删除菜单,打开信息接收
91       */
92      @Override
93      public void onMenuIsOpen(View view) {
94          mMenu=(SlidingButtonView) view;
95      }
96      /**
97       * 滑动或者点击Item监听
98       */
99      @Override
100     public void onDownOrMove(SlidingButtonView slidingButtonView) {
101         if(menuIsOpen()) {
102             if(mMenu !=slidingButtonView) {
103                 closeMenu();
104             }
105         }
106     }
107     /**
108      * 关闭菜单
109      */
110     public void closeMenu() {
111         mMenu.closeMenu();
112         mMenu=null;
113     }
114     /**
115      * 判断是否有菜单打开
116      */
117     public Boolean menuIsOpen() {
118         if(mMenu !=null) {
119             return true;
120         }
121         return false;
122     }
123     public interface IonSlidingViewClickListener {
124         void onItemClick(View view, int position);
125         void onDeleteBtnCilck(View view, int position);
126     }
127 }
```

【任务 12-4】 收藏新闻信息表

【任务分析】

由于"收藏"界面需要显示用户收藏的所有信息，因此需要在数据库中创建一个收藏信息的表，把用户收藏的信息保存到该表中，便于后续根据用户名查询该用户收藏的数据。

【任务实施】

（1）创建收藏新闻信息表。在 com.itheima.topline.sqlite 包的 SQLiteHelper 类中的 "public static final String CONSTELLATION = "constellation";//十二星座信息"语句下方添加如下代码：

```
//收藏新闻信息
public static final String COLLECTION_NEWS_INFO="collection_news_info";
```

在 SQLiteHelper 类中创建一个 COLLECTION_NEWS_INFO 收藏信息表，具体代码如下：

```
1    /**
2     * 创建收藏表
3     */
4    db.execSQL("CREATE TABLE  IF NOT EXISTS "+COLLECTION_NEWS_INFO+"( "
5            +"_id INTEGER PRIMARY KEY AUTOINCREMENT, "
6            +"id INTEGER, "                  //新闻 id
7            +"type INTEGER, "                //新闻类型
8            +"userName VARCHAR, "            //用户名
9            +"newsName VARCHAR, "            //新闻名称
10           +"newsTypeName VARCHAR,"         //新闻类型名称
11           +"img1 VARCHAR, "                //图片 1
12           +"img2 VARCHAR, "                //图片 2
13           +"img3 VARCHAR, "                //图片 3
14           +"newsUrl VARCHAR "              //新闻链接地址
15           +")");
```

在 SQLiteHelper 类的 onUpgrade() 方法中的 "db.execSQL("DROP TABLE IF EXISTS "+CONSTELLATION);"语句下方添加如下代码：

```
db.execSQL("DROP TABLE IF EXISTS "+COLLECTION_NEWS_INFO);
```

（2）保存收藏的数据到数据库中。由于收藏的新闻信息数据需要保存到数据库中，因此需要在 com.itheima.topline.utils 包中的 DBUtils 类中创建一个 saveCollectionNewsInfo() 方法保存收藏的新闻信息数据，具体代码如下所示：

```
1    /**
2     * 保存收藏信息
3     */
4    public void saveCollectionNewsInfo(NewsBean bean, String userName) {
```

```
5      ContentValues cv=new ContentValues();
6      cv.put("id", bean.getId());
7      cv.put("type", bean.getType());
8      cv.put("userName", userName);
9      cv.put("newsName", bean.getNewsName());
10     cv.put("newsTypeName", bean.getNewsTypeName());
11     cv.put("img1", bean.getImg1());
12     cv.put("img2", bean.getImg2());
13     cv.put("img3", bean.getImg3());
14     cv.put("newsUrl", bean.getNewsUrl());
15     db.insert(SQLiteHelper.COLLECTION_NEWS_INFO, null, cv);
16  }
```

（3）根据用户名从数据库获取收藏信息。由于"收藏"界面需要根据用户名（userName）查询该用户的收藏信息，因此需要在 com.itheima.topline.utils 包中的 DBUtils 类中创建一个 getCollectionNewsInfo() 方法获取收藏的信息数据，具体代码如下所示：

```
1   /**
2    * 获取收藏信息
3    */
4   public List<NewsBean>getCollectionNewsInfo(String userName) {
5     String sql="SELECT * FROM "+SQLiteHelper.COLLECTION_NEWS_INFO
6     +" WHERE  userName=? ";
7     Cursor cursor=db.rawQuery(sql, new String[]{userName});
8     List<NewsBean>newsList=new ArrayList<>();
9     NewsBean bean=null;
10    while (cursor.moveToNext()) {
11      bean=new NewsBean();
12      bean.setId(cursor.getInt(cursor.getColumnIndex("id")));
13      bean.setType(cursor.getInt(cursor.getColumnIndex("type")));
14      bean.setNewsName(cursor.getString(cursor.getColumnIndex("newsName")));
15      bean.setNewsTypeName(cursor.getString(cursor.getColumnIndex(
16                                                  "newsTypeName")));
17      bean.setImg1(cursor.getString(cursor.getColumnIndex("img1")));
18      bean.setImg2(cursor.getString(cursor.getColumnIndex("img2")));
19      bean.setImg3(cursor.getString(cursor.getColumnIndex("img3")));
20      bean.setNewsUrl(cursor.getString(cursor.getColumnIndex("newsUrl")));
21      newsList.add(bean);
22    }
23    cursor.close();
24    return newsList;
25  }
```

（4）判断新闻是否被收藏。"收藏"界面需要判断数据库中是否已经有该收藏数据，如果有，则删除本条收藏信息，重新保存一遍；如果没有，则直接保存收藏信息，因此需要在 com.itheima.topline.utils 包中的 DBUtils 类中创建一个 hasCollectionNewsInfo() 方法判断一条新闻信息是否被收藏，具体代码如下所示：

```
1   /**
2    * 判断一条新闻是否被收藏
3    */
4   public boolean hasCollectionNewsInfo(int id, int type, String userName) {
5       boolean hasNewsInfo=false;
6       String sql="SELECT * FROM "+SQLiteHelper.COLLECTION_NEWS_INFO
7           +" WHERE id=? AND type=? AND userName=?";
8       Cursor cursor=db.rawQuery(sql, new String[]{id+"", type+
9                                 "", userName+""});
10      if(cursor.moveToFirst()) {
11          hasNewsInfo=true;
12      }
13      cursor.close();
14      return hasNewsInfo;
15  }
```

（5）删除收藏信息。由于在删除收藏信息时也会删除数据库中保存的该条数据，因此需要在com.itheima.topline.utils包中的DBUtils类中创建一个delCollectionNewsInfo()方法删除某一条收藏的信息数据，具体代码如下所示：

```
1   /**
2    * 删除某一条收藏信息
3    */
4   public boolean delCollectionNewsInfo(int id, int type, String userName) {
5       boolean delSuccess=false;
6       if(hasCollectionNewsInfo(id, type, userName)) {
7           int row=db.delete(SQLiteHelper.COLLECTION_NEWS_INFO,
8           " id=? AND type=? AND userName=? ", new String[]{id+
9           "", type+"", userName});
10          if(row>0) {
11              delSuccess=true;
12          }
13      }
14      return delSuccess;
15  }
```

（6）修改"新闻详情"界面。由于点击"新闻详情"右上角的"收藏"图标才会收藏或取消收藏信息，因此需要在第7章中找到文件7-25，在该文件中的"private String position;"语句下方添加如下代码：

```
private boolean isCollection=false;
private DBUtils db;
private String userName;
```

在"if(bean==null) return;"语句下方添加如下代码：

```
db=DBUtils.getInstance(NewsDetailActivity.this);
```

在"newsUrl=bean.getNewsUrl();"语句下方添加如下代码：

```
userName=UtilsHelper.readLoginUserName(NewsDetailActivity.this);
```

在 init()方法中的"iv_collection.setVisibility(View.VISIBLE);"语句下方添加如下代码：

```
if(db.hasCollectionNewsInfo(bean.getId(),bean.getType(),userName)){
  iv_collection.setImageResource(R.drawable.collection_selected);
  isCollection=true;
}else{
  iv_collection.setImageResource(R.drawable.collection_normal);
  isCollection=false;
}
```

在 init()方法中的"收藏"按钮 iv_collection 的点击事件中添加如下代码：

```
if(UtilsHelper.readLoginStatus(NewsDetailActivity.this)) {
  if(isCollection) {
      iv_collection.setImageResource(R.drawable.collection_normal);
      isCollection=false;
      //删除保存到新闻收藏数据库中的数据
      db.delCollectionNewsInfo(bean.getId(), bean.getType(), userName);
      Toast.makeText(NewsDetailActivity.this,"取消收藏",Toast.LENGTH_SHORT).show();
      Intent data=new Intent();
      data.putExtra("position", position);
      setResult(RESULT_OK, data);
  } else {
      iv_collection.setImageResource(R.drawable.collection_selected);
      isCollection=true;
      //把该数据保存到新闻收藏数据库中
      db.saveCollectionNewsInfo(bean, userName);
      Toast.makeText(NewsDetailActivity.this, "收藏成功", Toast.LENGTH_SHORT).show();
  }
}else{
  Toast.makeText(NewsDetailActivity.this, "您还未登录,请先登录",Toast.LENGTH_SHORT).
}
```

【任务 12-5】 "收藏"界面逻辑代码

【任务分析】

在"收藏"界面中，当向左滑动"收藏"界面条目时会出现红色"删除"按钮，点击该按钮会删除当前收藏信息，同时也会删除数据库中对应的信息。点击"收藏"界面中的每个条目会跳转到对应的"新闻详情"界面。

【任务实施】

（1）获取界面控件。在 CollectionActivity 中创建界面控件的初始化方法 initView()，用于获取"收藏"界面所要用到的控件。

（2）设置"收藏"界面对应的适配器。在 CollectionActivity 中创建 setAdapter()方法，用于设置"收藏"界面对应的 CollectionAdapter 适配器。

（3）设置"收藏"界面 Item 的点击事件。在 CollectionActivity 中重写 onItemClick()方法，用于设置"收藏"界面 Item 的点击事件。

（4）接收回传数据。在 CollectionActivity 中重写 onActivityResult()方法，用于接收新闻详情界面取消收藏后传递过来的信息。具体代码如文件 12-7 所示。

【文件 12-7】 CollectionActivity.java

```
1   package com.itheima.topline.activity;
2   public class CollectionActivity extends AppCompatActivity implements
3   CollectionAdapter.IonSlidingViewClickListener {
4       private RecyclerView mRecyclerView;
5       private CollectionAdapter mAdapter;
6       private TextView tv_main_title, tv_back, tv_none;
7       private RelativeLayout rl_title_bar;
8       private DBUtils db;
9       private List<NewsBean> newsList;
10      private String userName;
11      @Override
12      protected void onCreate(Bundle savedInstanceState) {
13          super.onCreate(savedInstanceState);
14          setContentView(R.layout.activity_collection);
15          db=DBUtils.getInstance(CollectionActivity.this);
16          userName=UtilsHelper.readLoginUserName(CollectionActivity.this);
17          initView();
18          setAdapter();
19      }
20      private void initView() {
21          newsList=new ArrayList<>();
22          newsList=db.getCollectionNewsInfo(userName);
23          tv_main_title= (TextView) findViewById(R.id.tv_main_title);
24          tv_main_title.setText("收藏");
25          rl_title_bar=(RelativeLayout) findViewById(R.id.title_bar);
26          rl_title_bar.setBackgroundColor(getResources().getColor(R.color.
27          rdTextColorPress));
28          tv_back= (TextView) findViewById(R.id.tv_back);
29          tv_back.setVisibility(View.VISIBLE);
30          mRecyclerView= (RecyclerView) findViewById(R.id.rv_recyclerView);
31          tv_none= (TextView) findViewById(R.id.tv_none);
32          tv_back.setOnClickListener(new View.OnClickListener() {
33              @Override
34              public void onClick(View view) {
35                  CollectionActivity.this.finish();
36              }
37          });
38      }
```

```
39    private void setAdapter() {
40        mAdapter=new CollectionAdapter(this);
41        mRecyclerView.setLayoutManager(new LinearLayoutManager(this));
42        mRecyclerView.setAdapter(mAdapter);
43        mAdapter.setData(newsList);
44        if(newsList.size()==0) tv_none.setVisibility(View.VISIBLE);
45        mRecyclerView.setItemAnimator(new DefaultItemAnimator());
46    }
47    @Override
48    public void onItemClick(View view, int position) {
49        Intent intent=new Intent(CollectionActivity.this,
50                                 NewsDetailActivity.class);
51        intent.putExtra("newsBean", newsList.get(position));
52        intent.putExtra("position", position+"");
53        startActivityForResult(intent, 1);
54    }
55    @Override
56    public void onDeleteBtnCilck(View view, int position) {
57        mAdapter.removeData(position, tv_none, userName);
58    }
59    @Override
60    protected void onActivityResult(int requestCode, int resultCode, Intent data)
61    {
62        if(data !=null) {
63            String position=data.getStringExtra("position");
64            mAdapter.removeData(Integer.parseInt(position), tv_none, userName);
65        }
66    }
67 }
```

（5）修改"我"界面逻辑代码。由于点击"我"界面上的"收藏"条目时会跳转到"收藏"界面，因此需要在第 10 章中找到文件 10-10，在该文件的 onClick() 方法中的注释"//跳转到收藏界面"语句下方添加如下代码：

```
Intent collection=new Intent(getActivity(), CollectionActivity.class);
startActivity(collection);
```

12.2 设置

任务综述

"设置"界面主要包含"修改密码""设置密保""退出登录"功能。当用户点击"修改密码"时会跳转到"修改密码"界面，当用户点击"设置密保"时会跳转到"设置密保"界面，当点击"退出登录"时会退出当前登录的账号。

【任务 12-6】 "设置"界面

【任务分析】

根据任务综述可知,"设置"界面有三个功能,分别为修改密码、设置密保和退出登录,界面效果如图 12-3 所示。

图 12-3 "设置"界面

【任务实施】

(1) 创建"设置"界面。在 com.itheima.topline.activity 包中创建一个 Empty Activity 类,命名为 SettingActivity 并将布局文件名指定为 activity_setting。

(2) 放置界面控件。在布局文件中,放置 5 个 View 控件,用于显示 5 条灰色分隔线;2 个 ImageView 控件用于显示右边的箭头图片;3 个 TextView 控件用于显示界面文字(修改密码、设置密保和退出登录),具体代码如文件 12-8 所示。

【文件 12-8】 activity_setting.xml

```
1  <?xml version="1.0" encoding="utf-8"?>
2  <LinearLayout xmlns:android="http://schemas.android.com/apk/res/android"
3      android:layout_width="match_parent"
4      android:layout_height="match_parent"
5      android:background="@color/register_bg_color"
6      android:orientation="vertical">
7      <include layout="@layout/main_title_bar" />
8      <View
9          android:layout_width="fill_parent"
```

```xml
10        android:layout_height="1dp"
11        android:layout_marginTop="15dp"
12        android:background="#E3E3E3" />
13    <RelativeLayout
14        android:id="@+id/rl_modify_psw"
15        android:layout_width="fill_parent"
16        android:layout_height="50dp"
17        android:background="#F7F8F8"
18        android:gravity="center_vertical"
19        android:paddingLeft="15dp"
20        android:paddingRight="15dp">
21        <TextView
22            android:layout_width="wrap_content"
23            android:layout_height="wrap_content"
24            android:layout_centerVertical="true"
25            android:text="修改密码"
26            android:textColor="#A3A3A3"
27            android:textSize="14sp" />
28        <ImageView
29            android:layout_width="10dp"
30            android:layout_height="10dp"
31            android:layout_alignParentRight="true"
32            android:layout_centerVertical="true"
33            android:src="@drawable/iv_right_arrow" />
34    </RelativeLayout>
35    <View
36        android:layout_width="fill_parent"
37        android:layout_height="1dp"
38        android:background="#E3E3E3" />
39    <RelativeLayout
40        android:id="@+id/rl_security_setting"
41        android:layout_width="fill_parent"
42        android:layout_height="50dp"
43        android:background="#F7F8F8"
44        android:gravity="center_vertical"
45        android:paddingLeft="15dp"
46        android:paddingRight="15dp">
47        <TextView
48            android:layout_width="wrap_content"
49            android:layout_height="wrap_content"
50            android:layout_centerVertical="true"
51            android:text="设置密保"
52            android:textColor="#A3A3A3"
53            android:textSize="14sp" />
54        <ImageView
55            android:layout_width="10dp"
56            android:layout_height="10dp"
57            android:layout_alignParentRight="true"
58            android:layout_centerVertical="true"
```

```
59              android:src="@drawable/iv_right_arrow" />
60      </RelativeLayout>
61      <View
62          android:layout_width="fill_parent"
63          android:layout_height="1dp"
64          android:background="#E3E3E3" />
65      <View
66          android:layout_width="fill_parent"
67          android:layout_height="1dp"
68          android:layout_marginTop="15dp"
69          android:background="#E3E3E3" />
70      <RelativeLayout
71          android:id="@+id/rl_exit_login"
72          android:layout_width="fill_parent"
73          android:layout_height="50dp"
74          android:background="#F7F8F8"
75          android:gravity="center_vertical"
76          android:paddingLeft="15dp"
77          android:paddingRight="15dp">
78          <TextView
79              android:layout_width="wrap_content"
80              android:layout_height="wrap_content"
81              android:layout_centerVertical="true"
82              android:text="退出登录"
83              android:textColor="#A3A3A3"
84              android:textSize="14sp" />
85      </RelativeLayout>
86      <View
87          android:layout_width="fill_parent"
88          android:layout_height="1dp"
89          android:background="#E3E3E3" />
90  </LinearLayout>
```

【任务 12-7】 "设置"界面逻辑代码

【任务分析】

在"设置"界面中添加点击事件,当点击"修改密码"时跳转到"修改密码"界面,当点击"设置密保"时跳转到"设置密保"界面,当点击"退出登录"时清除登录状态和用户名,并且将退出的状态传递到"我"界面。

【任务实施】

(1) 获取界面控件。在 SettingActivity 中创建界面控件的初始化方法 init(),用于获取"设置"界面所要用到的控件以及设置后退键、修改密码、设置密保和退出登录的点击事件,具体代码如文件 12-9 所示。

【文件 12-9】 SettingActivity.java

```java
1   package com.itheima.topline.activity;
2   public class SettingActivity extends AppCompatActivity {
3       private TextView tv_main_title, tv_back;
4       private RelativeLayout rl_title_bar;
5       private RelativeLayout rl_modify_psw, rl_security_setting, rl_exit_login;
6       public static SettingActivity instance=null;
7       private SwipeBackLayout layout;
8       @Override
9       protected void onCreate(Bundle savedInstanceState) {
10          super.onCreate(savedInstanceState);
11          layout=(SwipeBackLayout) LayoutInflater.from(this).inflate(
12                  R.layout.base, null);
13          layout.attachToActivity(this);
14          setContentView(R.layout.activity_setting);
15          instance=this;
16          init();
17      }
18      private void init() {
19          tv_main_title=(TextView) findViewById(R.id.tv_main_title);
20          tv_main_title.setText("设置");
21          tv_back= (TextView) findViewById(R.id.tv_back);
22          rl_title_bar=(RelativeLayout) findViewById(R.id.title_bar);
23          rl_title_bar.setBackgroundColor(getResources().getColor(R.color.
24                  rdTextColorPress));
25          rl_modify_psw=(RelativeLayout) findViewById(R.id.rl_modify_psw);
26          rl_security_setting=(RelativeLayout) findViewById(
27                                              R.id.rl_security_setting);
28          rl_exit_login=(RelativeLayout) findViewById(R.id.rl_exit_login);
29          tv_back.setVisibility(View.VISIBLE);
30          tv_back.setOnClickListener(new View.OnClickListener() {
31              @Override
32              public void onClick(View v) {
33                  SettingActivity.this.finish();
34              }
35          });
36          //修改密码的点击事件
37          rl_modify_psw.setOnClickListener(new View.OnClickListener() {
38              @Override
39              public void onClick(View v) {
40                  //跳转到"修改密码"界面
41              }
42          });
43          //设置密保的点击事件
44          rl_security_setting.setOnClickListener(new View.OnClickListener() {
45              @Override
46              public void onClick(View v) {
47                  //跳转到"设置密保"界面
48              }
```

```
49              });
50              //退出登录的点击事件
51              rl_exit_login.setOnClickListener(new View.OnClickListener() {
52                  @Override
53                  public void onClick(View v) {
54                      Toast.makeText(SettingActivity.this, "退出登录成功",
55                          Toast.LENGTH_SHORT).show();
56                      //清除登录状态和登录时的用户名
57                      UtilsHelper.clearLoginStatus(SettingActivity.this);
58                      //退出登录成功后把退出成功的状态传递到 MeFragment 中
59                      Intent data=new Intent();
60                      data.putExtra("isLogin", false);
61                      setResult(RESULT_OK, data);
62                      SettingActivity.this.finish();
63                  }
64              });
65          }
66      }
```

（2）清除 SharedPreferences 中的登录状态和登录时的用户名。由于点击"退出登录"时需要清除 SharedPreferences 中的登录状态和登录时的用户名，因此需要在 com.itheima.topline.utils 包中的 UtilsHelper 类中创建一个 clearLoginStatus() 方法以实现此功能，具体代码如下所示：

```
/**
 * 清除 SharedPreferences 中的登录状态和登录时的用户名
 */
public static void clearLoginStatus(Context context){
    SharedPreferences sp=context.getSharedPreferences("loginInfo", Context.MODE_PRIVATE);
    SharedPreferences.Editor editor=sp.edit();              //获取编辑器
    editor.putBoolean("isLogin", false);                    //清除登录状态
    editor.putString("loginUserName", "");                  //清除用户名
    editor.commit();                                        //提交修改
}
```

（3）修改清单文件。由于"设置"界面向右滑动会关闭该界面，因此需要给该界面添加透明主题的样式，在清单文件的 SettingActivity 对应的 activity 标签中添加如下代码：

```
android:theme="@style/AppTheme.TransparentActivity"
```

（4）修改"我"界面逻辑代码。由于点击"我"界面上的设置条目时会跳转到"设置"界面，因此需要在第 10 章中找到文件 10-10，在该文件的 onClick() 方法中的注释"//跳转到设置界面"语句下方添加如下代码：

```
Intent settingIntent=new Intent(getActivity(), SettingActivity.class);
startActivityForResult(settingIntent, 1);
```

12.3 修改密码

任务综述

"修改密码"界面主要用于保证用户信息的安全性。修改密码需要输入一次原始密码和一次新密码,密码修改成功后需要把 SharedPreferences 中存储的旧密码修改成新密码。

【任务 12-8】 "修改密码"界面

【任务分析】

"修改密码"界面主要是让用户在必要时修改自己的原始密码,从而保证用户信息的安全性,界面效果如图 12-4 所示。

【任务实施】

(1) 创建"修改密码"界面。在 com.itheima.topline.activity 包中创建一个 Empty Activity 类,命名为 ModifyPswActivity 并将布局文件名指定为 activity_modify_psw。

(2) 放置界面控件。在布局文件中,放置两个 EditText 控件,分别用于输入原始密码与新密码,一个 Button 控件作为"保存"按钮,具体代码如文件 12-10 所示。

图 12-4 "修改密码"界面

【文件 12-10】 activity_modify_psw.xml

```
1   <?xml version="1.0" encoding="utf-8"?>
2   <LinearLayout xmlns:android="http://schemas.android.com/apk/res/android"
3       xmlns:app="http://schemas.android.com/apk/res-auto"
4       android:layout_width="match_parent"
5       android:layout_height="match_parent"
6       android:background="@color/register_bg_color"
7       android:orientation="vertical">
8       <include layout="@layout/main_title_bar" />
9       <android.support.v7.widget.CardView
10          android:layout_width="match_parent"
11          android:layout_height="wrap_content"
12          android:layout_marginLeft="25dp"
13          android:layout_marginRight="25dp"
14          android:layout_marginTop="35dp"
15          app:cardCornerRadius="5dp"
16          app:cardElevation="3dp"
17          app:cardPreventCornerOverlap="false"
```

```xml
18          app:cardUseCompatPadding="true">
19          <LinearLayout
20              android:layout_width="fill_parent"
21              android:layout_height="wrap_content"
22              android:orientation="vertical">
23              <EditText
24                  android:id="@+id/et_original_psw"
25                  android:layout_width="fill_parent"
26                  android:layout_height="48dp"
27                  android:layout_gravity="center_horizontal"
28                  android:background="@drawable/register_edittext_top_radius"
29                  android:drawableLeft="@drawable/psw_icon"
30                  android:drawablePadding="10dp"
31                  android:gravity="center_vertical"
32                  android:hint="请输入原始密码"
33                  android:inputType="textPassword"
34                  android:paddingLeft="8dp"
35                  android:textColor="#000000"
36                  android:textColorHint="@color/register_hint_text_color"
37                  android:textCursorDrawable="@null"
38                  android:textSize="14sp" />
39              <View
40                  android:layout_width="fill_parent"
41                  android:layout_height="1dp"
42                  android:background="@color/divider_line_color" />
43              <RelativeLayout
44                  android:layout_width="match_parent"
45                  android:layout_height="wrap_content">
46                  <EditText
47                      android:id="@+id/et_new_psw"
48                      android:layout_width="fill_parent"
49                      android:layout_height="48dp"
50                      android:layout_gravity="center_horizontal"
51                      android:background="@drawable/register_edittext_bottom_radius"
52                      android:drawableLeft="@drawable/psw_icon"
53                      android:drawablePadding="10dp"
54                      android:hint="请输入新密码"
55                      android:inputType="textPassword"
56                      android:paddingLeft="8dp"
57                      android:singleLine="true"
58                      android:textColor="#000000"
59                      android:textColorHint="#a3a3a3"
60                      android:textCursorDrawable="@null"
61                      android:textSize="14sp" />
62                  <ImageView
63                      android:id="@+id/iv_show_psw"
64                      android:layout_width="15dp"
65                      android:layout_height="48dp"
66                      android:layout_alignParentRight="true"
67                      android:layout_centerVertical="true"
```

```xml
68                    android:layout_marginRight="8dp"
69                    android:src="@drawable/hide_psw_icon" />
70            </RelativeLayout>
71        </LinearLayout>
72    </android.support.v7.widget.CardView>
73    <Button
74        android:id="@+id/btn_save"
75        android:layout_width="fill_parent"
76        android:layout_height="35dp"
77        android:layout_gravity="center_horizontal"
78        android:layout_marginLeft="25dp"
79        android:layout_marginRight="25dp"
80        android:layout_marginTop="15dp"
81        android:background="@drawable/register_btn_selector"
82        android:text="保存"
83        android:textColor="@android:color/white"
84        android:textSize="18sp" />
85 </LinearLayout>
```

【任务 12-9】 "修改密码"界面逻辑代码

【任务分析】

"修改密码"界面主要用于输入原始密码与新密码,输入的原始密码需要与从 SharedPreferences 中读取的原始密码一致,输入的新密码与原始密码不能相同。以上条件都符合之后,点击"保存"按钮将提示新密码设置成功,同时修改 SharedPreferences 中的原始密码。

【任务实施】

(1) 获取界面控件。在 ModifyPswActivity 中创建界面控件的初始化方法 init(),用于获取"修改密码"界面所要用到的控件以及设置后退键和"保存"按钮的点击事件。

(2) 修改 SharedPreferences 中的原始密码。由于在新密码设置成功时需要修改保存在 SharedPreferences 中的原始密码,因此需要创建 modifyPsw() 方法实现此功能,具体代码如文件 12-11 所示。

【文件 12-11】 ModifyPswActivity.java

```java
1  package com.itheima.topline.activity;
2  public class ModifyPswActivity extends AppCompatActivity{
3      private TextView tv_main_title, tv_back;
4      private Button btn_save;
5      private RelativeLayout rl_title_bar;
6      private EditText et_original_psw, et_new_psw;
7      private String originalPsw, newPsw;
8      private String userName;
9      private SwipeBackLayout layout;
10     private ImageView iv_show_psw;
```

```
11      private boolean isShowPsw=false;
12      @Override
13      protected void onCreate(Bundle savedInstanceState) {
14          super.onCreate(savedInstanceState);
15          layout=(SwipeBackLayout) LayoutInflater.from(this).inflate(
16          R.layout.base, null);
17          layout.attachToActivity(this);
18          setContentView(R.layout.activity_modify_psw);
19          init();
20          userName=UtilsHelper.readLoginUserName(this);
21      }
22      /**
23       * 获取界面控件并处理相关控件的点击事件
24       */
25      private void init() {
26          tv_main_title=(TextView) findViewById(R.id.tv_main_title);
27          tv_main_title.setText("修改密码");
28          tv_back=(TextView) findViewById(R.id.tv_back);
29          tv_back.setVisibility(View.VISIBLE);
30          rl_title_bar=(RelativeLayout) findViewById(R.id.title_bar);
31          rl_title_bar.setBackgroundColor(getResources().getColor(R.color.
32          rdTextColorPress));
33          et_original_psw=(EditText) findViewById(R.id.et_original_psw);
34          et_new_psw=(EditText) findViewById(R.id.et_new_psw);
35          iv_show_psw=(ImageView) findViewById(R.id.iv_show_psw);
36          btn_save=(Button) findViewById(R.id.btn_save);
37          tv_back.setOnClickListener(new View.OnClickListener() {
38              @Override
39              public void onClick(View v) {
40                  ModifyPswActivity.this.finish();
41              }
42          });
43          iv_show_psw.setOnClickListener(new View.OnClickListener() {
44              @Override
45              public void onClick(View view) {
46                  newPsw=et_new_psw.getText().toString();
47                  if(isShowPsw) {
48                      iv_show_psw.setImageResource(R.drawable.hide_psw_icon);
49                      //隐藏密码
50                      et_new_psw.setTransformationMethod(PasswordTransformationMethod.
51                      getInstance());
52                      isShowPsw=false;
53                      if(newPsw !=null) {
54                          et_new_psw.setSelection(newPsw.length());
55                      }
56                  } else {
57                      iv_show_psw.setImageResource(R.drawable.show_psw_icon);
58                      //显示密码
```

```java
59                  et_new_psw.setTransformationMethod(
60                          HideReturnsTransformationMethod.getInstance());
61                  isShowPsw=true;
62                  if(newPsw !=null) {
63                      et_new_psw.setSelection(newPsw.length());
64                  }
65              }
66          }
67      });
68      //"保存"按钮的点击事件
69      btn_save.setOnClickListener(new View.OnClickListener() {
70          @Override
71          public void onClick(View v) {
72              getEditString();
73              if(TextUtils.isEmpty(originalPsw)) {
74                  Toast.makeText(ModifyPswActivity.this, "请输入原始密码",
75                          Toast.LENGTH_SHORT).show();
76                  return;
77              } else if(!MD5Utils.md5(originalPsw).equals(readPsw())) {
78                  Toast.makeText(ModifyPswActivity.this, "输入的密码与原始密码不一致",
79                          Toast.LENGTH_SHORT).show();
80                  return;
81              } else if(MD5Utils.md5(newPsw).equals(readPsw())) {
82                  Toast.makeText(ModifyPswActivity.this, "输入的新密码与原始密码不能一致",
83                          Toast.LENGTH_SHORT).show();
84                  return;
85              } else if(TextUtils.isEmpty(newPsw)) {
86                  Toast.makeText(ModifyPswActivity.this, "请输入新密码",
87                          Toast.LENGTH_SHORT).show();
88                  return;
89              } else {
90                  Toast.makeText(ModifyPswActivity.this, "新密码设置成功",
91                          Toast.LENGTH_SHORT).show();
92                  //修改登录成功时保存在SharedPreferences中的密码
93                  modifyPsw(newPsw);
94                  Intent intent=new Intent(ModifyPswActivity.this,
95                          LoginActivity.class);
96                  startActivity(intent);
97                  SettingActivity.instance.finish();      //关闭设置界面
98                  ModifyPswActivity.this.finish();        //关闭本界面
99              }
100         }
101     });
102 }
103 /**
104  * 获取控件上的字符串
105  */
106 private void getEditString() {
```

```
107          originalPsw=et_original_psw.getText().toString().trim();
108          newPsw=et_new_psw.getText().toString().trim();
109      }
110      /**
111       * 修改登录成功时保存在 SharedPreferences 中的密码
112       */
113      private void modifyPsw(String newPsw) {
114          String md5Psw=MD5Utils.md5(newPsw);              //把密码用 MD5 加密
115          SharedPreferences sp=getSharedPreferences("loginInfo", MODE_PRIVATE);
116          SharedPreferences.Editor editor=sp.edit();        //获取编辑器
117          editor.putString(userName, md5Psw);               //保存新密码
118          editor.commit();                                  //提交修改
119      }
120      /**
121       * 从 SharedPreferences 中读取原始密码
122       */
123      private String readPsw() {
124          SharedPreferences sp=getSharedPreferences("loginInfo", MODE_PRIVATE);
125          String spPsw=sp.getString(userName, "");
126          return spPsw;
127      }
128  }
```

（3）修改清单文件。由于"修改密码"界面向右滑动会关闭该界面，因此需要给该界面添加透明主题的样式。在清单文件的 ModifyPswActivity 对应的 activity 标签中添加如下代码：

```
android:theme="@style/AppTheme.TransparentActivity"
```

（4）修改"设置"界面逻辑代码。由于点击"设置"界面上的"修改密码"条目会跳转到"修改密码"界面，因此需要找到文件 12-9 中的 init()方法，在该方法中的注释"//跳转到修改密码的界面"语句下方添加如下代码：

```
Intent intent=new Intent(SettingActivity.this, ModifyPswActivity.class);
startActivity(intent);
```

12.4 设置密保

任务综述

根据功能展示可知，"设置密保"界面和"找回密码"界面基本相同，同时，两个界面的代码逻辑也十分相似，因此这两个界面可以使用同一个布局文件，也可以使用同一个 Activity 处理逻辑代码。设置密保主要是将当前用户输入的姓名作为密保，找回密码是根据用户输入的用户名和密保姓名将该用户的密码重置为初始密码 123456（由于之前保存的密码是经过 MD5 加密的，且 MD5 是不可逆的，所以之前的密码无法获取明文）。

【任务 12-10】 "设置密保"界面

【任务分析】

"设置密保"界面主要用于输入要设为密保的姓名,"找回密码"界面可以根据用户当前输入的用户名和设为密保的姓名找回密码,界面效果如图 12-5 所示。

图 12-5 "设置密保"和"找回密码"界面

【任务实施】

(1) 创建"设置密保"和"找回密码"界面。在 com.itheima.topline.activity 包中创建一个 Empty Activity 类,命名为 FindPswActivity 并将布局文件名指定为 activity_find_psw。

(2) 放置界面控件。在布局文件中,放置两个 EditText 控件,用于输入用户名和姓名; 3 个 TextView 控件,一个用于显示密码(此控件暂时隐藏),其余两个分别用于显示"您的用户名是?"和"您的姓名是?"文字;一个 Button 控件作为"验证"按钮,具体代码如文件 12-12 所示。

【文件 12-12】 activity_find_psw.xml

```
1   <?xml version="1.0" encoding="utf-8"?>
2   <LinearLayout xmlns:android="http://schemas.android.com/apk/res/android"
3       android:layout_width="match_parent"
4       android:layout_height="match_parent"
5       android:background="@color/register_bg_color"
6       android:orientation="vertical">
7       <include layout="@layout/main_title_bar" />
8       <TextView
```

```
9            android:id="@+id/tv_user_name"
10           android:layout_width="fill_parent"
11           android:layout_height="wrap_content"
12           android:layout_marginLeft="25dp"
13           android:layout_marginRight="25dp"
14           android:layout_marginTop="35dp"
15           android:text="您的用户名是?"
16           android:textColor="@color/constellation_info_color"
17           android:textSize="14sp"
18           android:visibility="gone" />
19       <EditText
20           android:id="@+id/et_user_name"
21           android:layout_width="fill_parent"
22           android:layout_height="40dp"
23           android:layout_marginLeft="25dp"
24           android:layout_marginRight="25dp"
25           android:layout_marginTop="10dp"
26           android:background="@drawable/find_psw_radius_bg"
27           android:hint="请输入您的用户名"
28           android:paddingLeft="8dp"
29           android:singleLine="true"
30           android:textColor="#000000"
31           android:textColorHint="#a3a3a3"
32           android:textCursorDrawable="@null"
33           android:textSize="14sp"
34           android:visibility="gone" />
35       <TextView
36           android:layout_width="fill_parent"
37           android:layout_height="wrap_content"
38           android:layout_marginLeft="25dp"
39           android:layout_marginRight="25dp"
40           android:layout_marginTop="15dp"
41           android:text="您的姓名是?"
42           android:textColor="@color/constellation_info_color"
43           android:textSize="14sp" />
44       <EditText
45           android:id="@+id/et_validate_name"
46           android:layout_width="fill_parent"
47           android:layout_height="40dp"
48           android:layout_marginLeft="25dp"
49           android:layout_marginRight="25dp"
50           android:layout_marginTop="10dp"
51           android:background="@drawable/find_psw_radius_bg"
52           android:hint="请输入要验证的姓名"
53           android:paddingLeft="8dp"
54           android:singleLine="true"
55           android:textColor="#000000"
56           android:textColorHint="#a3a3a3"
57           android:textCursorDrawable="@null"
58           android:textSize="14sp" />
```

```
59    <TextView
60        android:id="@+id/tv_reset_psw"
61        android:layout_width="fill_parent"
62        android:layout_height="wrap_content"
63        android:layout_marginLeft="25dp"
64        android:layout_marginRight="25dp"
65        android:layout_marginTop="10dp"
66        android:gravity="center_vertical"
67        android:textColor="@android:color/holo_red_light"
68        android:textSize="14sp"
69        android:visibility="gone" />
70    <Button
71        android:id="@+id/btn_validate"
72        android:layout_width="fill_parent"
73        android:layout_height="35dp"
74        android:layout_gravity="center_horizontal"
75        android:layout_marginLeft="25dp"
76        android:layout_marginRight="25dp"
77        android:layout_marginTop="15dp"
78        android:background="@drawable/register_btn_selector"
79        android:text="验证"
80        android:textColor="@android:color/white"
81        android:textSize="16sp" />
82 </LinearLayout>
```

（3）创建 EditText 控件的背景。由于"设置密保"界面的输入框的四个角都是椭圆形，因此需要在 res/drawable 文件夹中创建一个 find_psw_radius_bg.xml 文件进行设置，具体代码如文件 12-13 所示。

【文件 12-13】 find_psw_radius_bg.xml

```
1 <?xml version="1.0" encoding="utf-8"?>
2 <shape xmlns:android="http://schemas.android.com/apk/res/android"
3     android:shape="rectangle">
4     <corners android:radius="5dp" />
5     <solid android:color="#FFFFFF" />
6     <stroke
7         android:width="1dp"
8         android:color="@color/register_btn_color" />
9 </shape>
```

【任务 12-11】 "设置密保"界面逻辑代码

【任务分析】

根据任务综述可知，"设置密保"界面和"找回密码"界面使用同一个 Activity，该 Activity 主要是根据从"设置"界面和"登录"界面传递过来的 from 参数的值判断要跳转到哪个界面，若值为 security，则处理的是"设置密保"界面，否则处理的就是"找回密码"界面。"设置密保"界面的逻辑主要是将用户输入的姓名保存到 SharedPreferences 中，"找回密码"

界面的逻辑主要是将 SharedPreferences 中用户名对应的原始密码修改为 123456。

【任务实施】

（1）获取界面控件。在 FindPswActivity 中创建界面控件的初始化方法 init()，用于获取"修改密码"界面所要用到的控件以及设置后退键、"保存"按钮的点击事件。

（2）保存密保。由于"设置密保"界面需要将用户输入的姓名保存到 SharedPreferences 中，因此需要创建 saveSecurity() 方法进行保存。

（3）保存初始化密码到 SharedPreferences 中。在"找回密码"界面创建 isExistUserName() 方法判断用户输入的用户名是否存在，若存在，则创建 readSecurity() 方法获取此用户之前设置过的密保，若用户输入的密保和从 SharedPreferences 中获取的密保一致，则创建 savePsw() 方法将此用户原来的密码保存为 123456（由于原来的密码不能获取明文，因此将此账户的密码重置为初始密码 123456），具体代码如文件 12-14 所示。

【文件 12-14】 FindPswActivity.java

```
1  package com.itheima.topline.activity;
2  public class FindPswActivity extends AppCompatActivity{
3      private EditText et_validate_name, et_user_name;
4      private Button btn_validate;
5      private TextView tv_main_title;
6      private TextView tv_back;
7      //from 为 security 时表示是从"设置密保"界面跳转过来的,否则就是从"登录"界面跳转过来的
8      private String from;
9      private TextView tv_reset_psw, tv_user_name;
10     private SwipeBackLayout layout;
11     private RelativeLayout rl_title_bar;
12     @Override
13     protected void onCreate(Bundle savedInstanceState) {
14         super.onCreate(savedInstanceState);
15         layout= (SwipeBackLayout) LayoutInflater.from(this).inflate(
16                 R.layout.base, null);
17         layout.attachToActivity(this);
18         setContentView(R.layout.activity_find_psw);
19         //获取从"登录"界面和"设置"界面传递过来的数据
20         from=getIntent().getStringExtra("from");
21         init();
22     }
23     /**
24      * 获取界面控件及处理相应控件的点击事件
25      */
26     private void init() {
27         tv_main_title= (TextView) findViewById(R.id.tv_main_title);
28         tv_back= (TextView) findViewById(R.id.tv_back);
29         tv_back.setVisibility(View.VISIBLE);
30         rl_title_bar= (RelativeLayout) findViewById(R.id.title_bar);
31         rl_title_bar.setBackgroundColor(getResources().getColor(R.color.
32                 rdTextColorPress));
33         et_validate_name= (EditText) findViewById(R.id.et_validate_name);
```

```java
34        btn_validate= (Button) findViewById(R.id.btn_validate);
35        tv_reset_psw= (TextView) findViewById(R.id.tv_reset_psw);
36        et_user_name= (EditText) findViewById(R.id.et_user_name);
37        tv_user_name= (TextView) findViewById(R.id.tv_user_name);
38        if("security".equals(from)) {
39            tv_main_title.setText("设置密保");
40        } else {
41            tv_main_title.setText("找回密码");
42            tv_user_name.setVisibility(View.VISIBLE);
43            et_user_name.setVisibility(View.VISIBLE);
44        }
45        tv_back.setOnClickListener(new View.OnClickListener() {
46            @Override
47            public void onClick(View v) {
48                FindPswActivity.this.finish();
49            }
50        });
51        btn_validate.setOnClickListener(new View.OnClickListener() {
52            @Override
53            public void onClick(View v) {
54                String validateName=et_validate_name.getText().toString().trim();
55                if("security".equals(from)) {           //设置密保
56                    if(TextUtils.isEmpty(validateName)) {
57                        Toast.makeText(FindPswActivity.this,"请输入要验证的姓名",
58                                Toast.LENGTH_SHORT).show();
59                        return;
60                    } else {
61                        Toast.makeText(FindPswActivity.this,"密保设置成功",
62                                Toast.LENGTH_SHORT).show();
63                        //保存密保到 SharedPreferences 中
64                        saveSecurity(validateName);
65                        FindPswActivity.this.finish();
66                    }
67                } else {         //找回密码
68                    String userName=et_user_name.getText().toString().trim();
69                    String sp_security=readSecurity(userName);
70                    if(TextUtils.isEmpty(userName)) {
71                        Toast.makeText(FindPswActivity.this,"请输入您的用户名",
72                                Toast.LENGTH_SHORT).show();
73                        return;
74                    } else if(!isExistUserName(userName)) {
75                        Toast.makeText(FindPswActivity.this,"您输入的用户名不存在",
76                                Toast.LENGTH_SHORT).show();
77                        return;
78                    } else if(TextUtils.isEmpty(validateName)) {
79                        Toast.makeText(FindPswActivity.this,"请输入要验证的姓名",
80                                Toast.LENGTH_SHORT).show();
81                        return;
82                    }
83                    if(!validateName.equals(sp_security)) {
```

```java
84                     Toast.makeText(FindPswActivity.this,"输入的密保不正确",
85                             Toast.LENGTH_SHORT).show();
86                     return;
87                 } else {
88                     //输入的密保正确,重新给用户设置一个密码
89                     tv_reset_psw.setVisibility(View.VISIBLE);
90                     tv_reset_psw.setText("初始密码:123456");
91                     savePsw(userName);
92                 }
93             }
94         }
95     });
96 }
97 /**
98  * 保存初始化的密码
99  */
100 private void savePsw(String userName) {
101     String md5Psw=MD5Utils.md5("123456");          //把密码用MD5加密
102     SharedPreferences sp=getSharedPreferences("loginInfo", MODE_PRIVATE);
103     SharedPreferences.Editor editor=sp.edit();     //获取编辑器
104     editor.putString(userName, md5Psw);
105     editor.commit();                               //提交修改
106 }
107 /**
108  * 保存密保到SharedPreferences中
109  */
110 private void saveSecurity(String validateName) {
111     SharedPreferences sp=getSharedPreferences("loginInfo", MODE_PRIVATE);
112     SharedPreferences.Editor editor=sp.edit();     //获取编辑器
113     editor.putString(UtilsHelper.readLoginUserName(this)+"_security",
114             validateName);                         //存入用户对应的密保
115     editor.commit();                               //提交修改
116 }
117 /**
118  * 从SharedPreferences中读取密保
119  */
120 private String readSecurity(String userName) {
121     SharedPreferences sp=getSharedPreferences("loginInfo",
122                                                 Context.MODE_PRIVATE);
123     String security=sp.getString(userName+"_security", "");
124     return security;
125 }
126 /**
127  * 从SharedPreferences中根据用户输入的用户名判断是否有此用户
128  */
129 private boolean isExistUserName(String userName) {
130     boolean hasUserName=false;
131     SharedPreferences sp=getSharedPreferences("loginInfo", MODE_PRIVATE);
132     String spPsw=sp.getString(userName, "");
133     if(!TextUtils.isEmpty(spPsw)) {
```

```
134              hasUserName=true;
135          }
136          return hasUserName;
137      }
138 }
```

(4)修改清单文件。由于"设置密保"界面向右滑动会关闭该界面,因此需要给该界面添加透明主题的样式,在清单文件的 FindPswActivity 对应的 activity 标签中添加如下代码:

```
android:theme="@style/AppTheme.TransparentActivity"
```

(5)"修改登录"界面。由于点击登录界面上的"忘记密码"文字会跳转到"找回密码"界面,因此需要在第 10 章中找到文件 10-19 中的 onClick()方法,在该方法中的"case R.id.tv_forget_psw:"语句下方添加如下代码:

```
Intent forget=new Intent(LoginActivity.this,FindPswActivity.class);
startActivity(forget);
```

(6)"修改设置"界面逻辑代码。由于点击"设置"界面上的"设置密保"条目会跳转到"设置密保"界面,因此需要找到文件 12-9 中的 init()方法,在该方法中的注释"//跳转到设置密保界面"语句下方添加如下代码:

```
Intent intent=new Intent(SettingActivity.this, FindPswActivity.class);
intent.putExtra("from", "security");
startActivity(intent);
```

12.5 本章小结

本章主要讲解了设置模块,包括收藏、设置、修改密码、设置密保等功能。读者通过本章的学习,可以掌握界面的搭建过程以及简单的界面开发与数据存储。

【思考题】

1. 如何实现收藏功能?
2. 如何设置密保?

第 13 章 项目上线

学习目标
- 掌握代码混淆方式以及项目打包流程,实现项目打包
- 掌握第三方加固软件的使用,使用该软件对项目进行加固
- 掌握项目发布到市场的流程,能够将头条项目上传到应用市场

当应用程序开发完成之后,需要将程序放到应用市场中供用户使用。在上传到应用市场之前,需要对程序代码进行混淆、打包、加固等,以提高程序的安全性。所有企业的项目都必须经历这一步,因此读者需要认真学习。本章将针对项目上线进行详细讲解。

13.1 代码混淆

思政材料 13

为了防止自己开发的程序被别人反编译并保护自己的劳动成果,一般情况下需要对程序进行代码混淆。代码混淆(也称花指令)是指保持程序功能不变,将程序代码转换成一种难以阅读和理解的形式。代码混淆为应用程序增加了一层保护措施,但是并不能完全防止程序被反编译。下面将对代码混淆进行详细讲解。

13.1.1 修改 build.gradle 文件

由于需要开启项目的混淆设置,因此需要在 build.gradle 文件的 buildTypes 中添加相关属性,具体代码如下所示:

```
1  buildTypes {
2      release {
3          minifyEnabled true
4          shrinkResources true
5          proguardFiles getDefaultProguardFile('proguard-android.txt'),
6                  'proguard-rules.pro'
7      }
8  }
```

在上述代码中,minifyEnabled 用于设置是否开启混淆,默认情况下为 false,需要开启混淆时设置为 true。shrinkResources 属性用于去除无用的 resource 文件。proguardFiles getDefaultProguardFile 用于加载混淆的配置文件,配置文件中含有混淆的相关规则。

13.1.2 编写 proguard-rules.pro 文件

代码混淆需要指定混淆规则,例如指定代码压缩级别、混淆时采用的算法、排除混淆的

类等,这些混淆规则是在 proguard-rules.pro 文件中编写的,具体代码如文件 13-1 所示。

【文件 13-1】 proguard-rules.pro

```
1   -ignorewarnings                                              #抑制警告
2   -keep class com.itheima.topline.bean.**{*;}                  #保持实体类不被混淆
3   -optimizationpasses 5                                        #指定代码的压缩级别
4   -dontusemixedcaseclassnames                                  #是否使用大小写混合
5   -dontpreverify                                               #混淆时是否做预校验
6   -verbose                                                     #混淆时是否记录日志
7   #指定混淆时采用的算法
8   -optimizations !code/simplification/arithmetic,!field/*,!class/merging/*
9   #百度地图混淆
10  -keep class com.amap.api.**{*;}
11  -keep class com.autonavi.amap.mapcore.**{*;}
12  #对于继承 Android 的四大组件等系统类,保持不被混淆
13  -keep public class * extends android.app.Activity
14  -keep public class * extends android.app.Application
15  -keep public class * extends android.app.Service
16  -keep public class * extends android.content.BroadcastReceiver
17  -keep public class * extends android.content.ContentProvider
18  -keep public class * extends android.app.backup.BackupAgentHelper
19  -keep public class * extends android.preference.Preference
20  -keep public class * extends android.view.View
21  -keep public class com.android.vending.licensing.ILicensingService
22  -keep class android.support.**{*;}
23  -keepclasseswithmembernames class * { #保持 native 方法不被混淆
24      native <methods>;
25  }
26  #保持自定义控件类不被混淆
27  -keepclassmembers class * extends android.app.Activity {
28      public void *(android.view.View);
29  }
30  #保持枚举类 enum 不被混淆
31  -keepclassmembers enum * {
32      public static **[] values();
33      public static ** valueOf(java.lang.String);
34  }
35  #保持 Parcelable 的类不被混淆
36  -keep class * implements android.os.Parcelable {
37      public static final android.os.Parcelable$Creator *;
38  }
39  #保持继承自 View 对象中的 set/get 方法以及初始化方法的方法名不被混淆
40  -keep public class * extends android.view.View{
41      *** get*();
42      void set*(***);
43      public <init>(android.content.Context);
44      public <init>(android.content.Context, android.util.AttributeSet);
45      public <init>(android.content.Context, android.util.AttributeSet, int);
46  }
47  #对所有类的初始化方法的方法名不进行混淆
```

```
48  -keepclasseswithmembers class * {
49      public<init>(android.content.Context, android.util.AttributeSet);
50      public<init>(android.content.Context, android.util.AttributeSet, int);
51  }
52  #保持 Serializable 序列化的类不被混淆
53  -keepclassmembers class * implements java.io.Serializable {
54      static final long serialVersionUID;
55      private static final java.io.ObjectStreamField[] serialPersistentFields;
56      private void writeObject(java.io.ObjectOutputStream);
57      private void readObject(java.io.ObjectInputStream);
58      java.lang.Object writeReplace();
59      java.lang.Object readResolve();
60  }
61  #对于 R(资源)下的所有类及其方法,都不能被混淆
62  -keep class *.*.R$* {
63   *;
64  }
65  #对于带有回调函数 onXXEvent 的,不能被混淆
66  -keepclassmembers class * {
67      void *(**On*Event);
68  }
69  #Webview 混淆的处理
70  -keepclassmembers class fqcn.of.javascript.interface.for.Webview {
71     public *;
72  }
73  -keepclassmembers class * extends android.webkit.WebViewClient {
74      public void *(android.webkit.WebView, java.lang.String,
75                                            android.graphics.Bitmap);
76      public boolean *(android.webkit.WebView, java.lang.String);
77  }
78  -keepclassmembers class * extends android.webkit.WebViewClient {
79      public void *(android.webkit.WebView, jav.lang.String);
80  }
```

从上述代码可以看出,在 proguard-rules.pro 文件中需要指定混淆时的一些属性,如代码压缩级别、是否使用大小写混合、混淆时的算法等。同时,在文件中还需要指定排除哪些类不被混淆,如 Activity 相关类、四大组件、自定义控件等,这些类若被混淆,则在程序打包后运行时将无法找到该类,因此需要将这些内容保持原样,不进行混淆。

13.2　项目打包

项目开发完成后,如果要发布到互联网上供别人使用,就需要将自己的程序打包成正式的 Android 安装包文件,简称 APK,其扩展名为 apk。下面针对 Android 程序打包过程进行详细讲解。

首先,在菜单栏中单击 Build→Generate Signed APK,进入 Generate Signed APK 界面。在该界面中单击 Create new 按钮,进入 New Key Store 界面,创建一个新的证书,如图 13-1 所示。

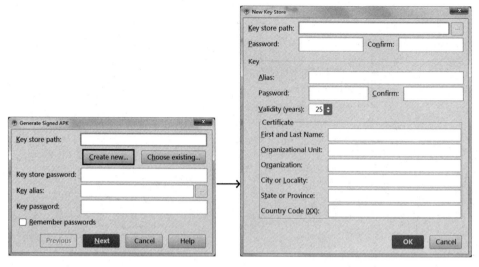

图 13-1　创建新的证书

在图 13-1 中，单击 Key store path 项之后的"…"按钮，进入 Choose keystore file 界面，选择证书存放路径，并在下方的 File name 中填写证书名称，如图 13-2 所示。

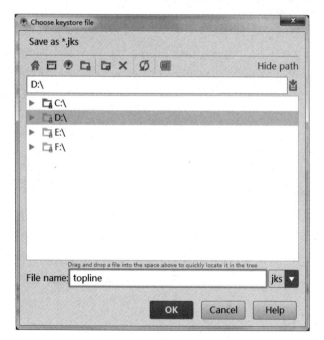

图 13-2　Choose keystore file 界面

在图 13-2 中，单击 OK 按钮。此时会返回到 New Key Store 界面，然后填写相关信息，如图 13-3 所示。

在图 13-3 中，信息填写完毕之后，单击 OK 按钮，返回到 Generate Signed APK 界面。然后单击 Next 按钮，选择 APK 文件的路径以及构建类型，如图 13-4 所示。

在图 13-4 中，APK Destination Folder 表示 APK 文件路径，Build Type 表示构建类型

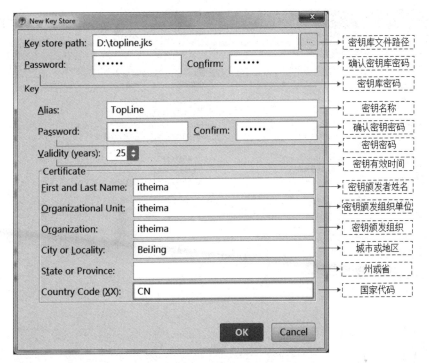

图 13-3 New Key Store 界面

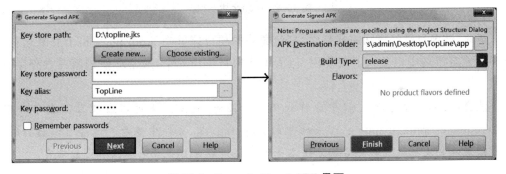

图 13-4 Generate Signed APK 界面

(有两种：Debug 和 Release。Debug 通常称为调试版本，包含调试信息，并且不进行任何优化，便于程序调试。Release 称为发布版本，往往进行了各种优化，以便用户更好地使用)。

此处选择 release，然后单击 Finish 按钮，Android Studio 的右上角会弹出一个显示 Generate Signed APK 的窗口，如图 13-5 所示。

图 13-5 APK(s) generated successfully 界面

在图 13-5 中，单击 Show in Explorer 文字，即可查看生成的 APK 文件，如图 13-6 所示。

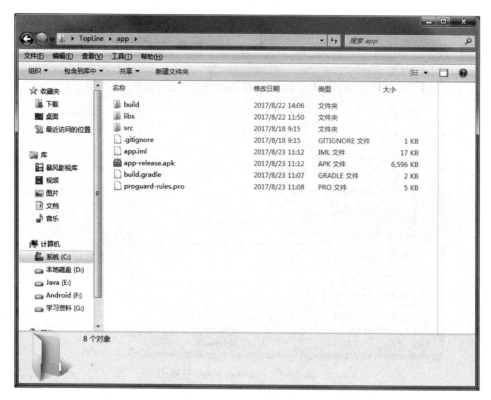

图 13-6 成功生成 APK

至此,该项目已经成功完成打包,打包成功的项目能够在 Android 手机上安装运行,也能够上传至市场中供他人下载使用,但为了使项目更加安全,通常会使用第三方程序进行加固。

注意:在项目打包的过程中会将代码进行混淆,混淆结果可以在项目所在路径下的\app\build\outputs\mapping\release 中的 mapping.txt 文件中查看。读者可以自行验证,打开该文件会发现项目的类名和方法名等已经混淆成 a、b、c、d 等难以解读的内容。

13.3 项目加固

在实际开发中,为了增强项目的安全性,增加代码的健硕程度,会根据项目需求使用第三方的加固软件对项目进行加固(加密)。下面对第三方加固软件"360 加固助手"进行详细讲解。

1. 下载加固助手

首先,进入 360 加固的网站首页(http://jiagu.360.cn/),在注册登录后,找到"加固助手"的下载页面(http://jiagu.360.cn/qcmshtml/details.html#helper)。选择与操作系统相对应的软件进行下载,本文以 Windows 操作系统为例。下载完成后进行解压,然后打开 360 加固助手,如图 13-7 所示。

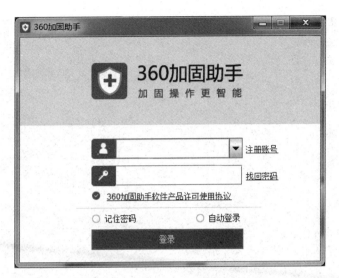

图 13-7 登录

输入账号和密码后单击"登录"按钮,进入账号信息填写界面,如图 13-8 所示。

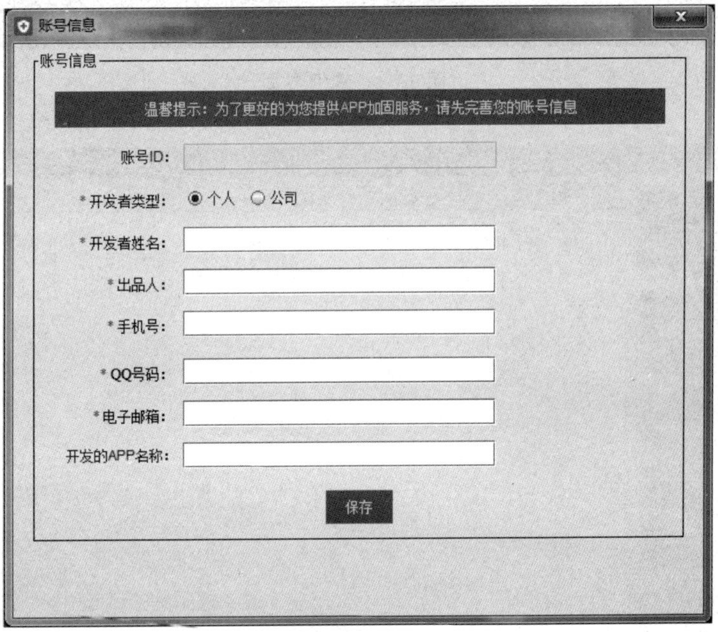

图 13-8 账号信息

填写完成账号信息后单击"保存"按钮,进入程序的欢迎界面,如图 13-9 所示。
在欢迎界面可以选择"查看新手引导"进行学习,读者可以自行查看,本书不再演示。

2．配置信息

在欢迎界面中,单击"开始使用"按钮,进入加固助手界面,如图 13-10 所示。

图 13-9 欢迎界面

图 13-10 加固助手

首先单击"配置信息"按钮进入"配置信息"界面,在"签名配置"选项卡中勾选"启用自动签名"即可添加本地的 keystore 签名文件,选择文件路径(D:\topline.jks)并输入 keystore 密码,如图 13-11 所示。

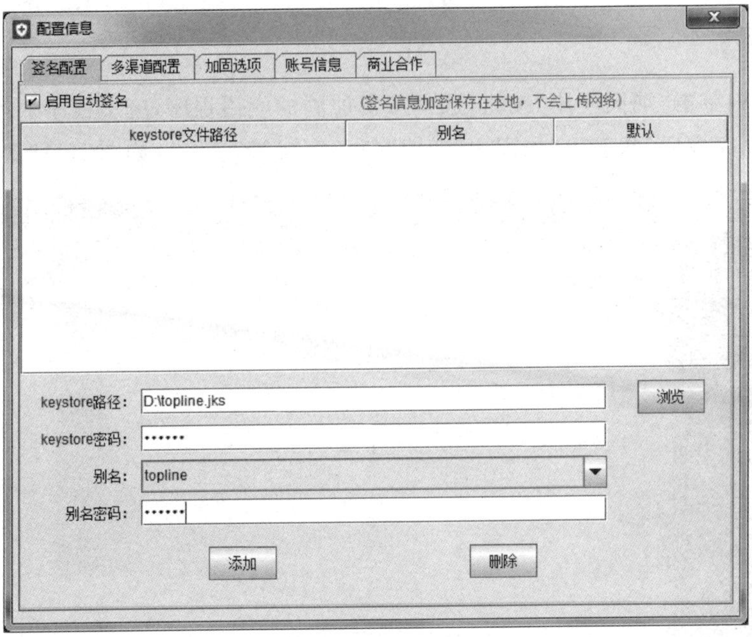

图 13-11　填写配置信息

在图 13-11 中填写配置信息,单击"添加"按钮即可完成配置信息的添加,如图 13-12 所示。

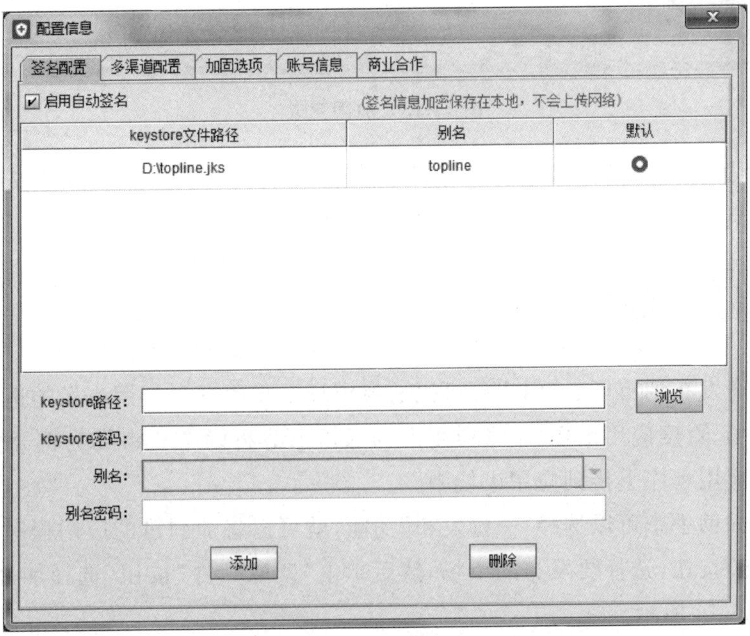

图 13-12　添加配置信息

在图 13-12 中，单击"多渠道配置"选项卡可以进行多渠道打包；单击"加固选项"选项卡可以选择文件的输出路径以及一些增强服务；单击"账号信息"选项卡可以查看账号的相关内容。填写完毕后，关闭该窗口即可完成配置。

3．加固应用

在主界面中单击"加固应用"按钮，选择需要加固的应用程序，应用程序上传后会处于加固状态，当加固完成后，状态由"加固中"变成"任务完成_已签名"，如图 13-13 所示。

图 13-13　加固完成

至此，使用第三方工具加固应用程序全部完成，完成加密后的应用程序安全性更高，将应用程序上传至应用市场即可供其他用户使用。

13.4　项目发布

在应用程序发布到市场后，用户便可以通过市场下载程序。应用市场的选择也有很多，如 360 应用市场、百度应用市场、小米应用市场等。本节将以 360 市场为例，将详细讲解如何将加固后的应用程序上传到应用市场中。

在 360 加固助手中可以选择"一键发布"功能，就可以将加固过的应用程序上传至市场。单击"一键发布"按钮，选择要发布的市场，然后单击"读取文件"按钮，选择加固完成后的应用程序，如图 13-14 所示。

文件读取完成后，单击"填写发布资料"按钮，进入资料填写界面，如图 13-15 所示。

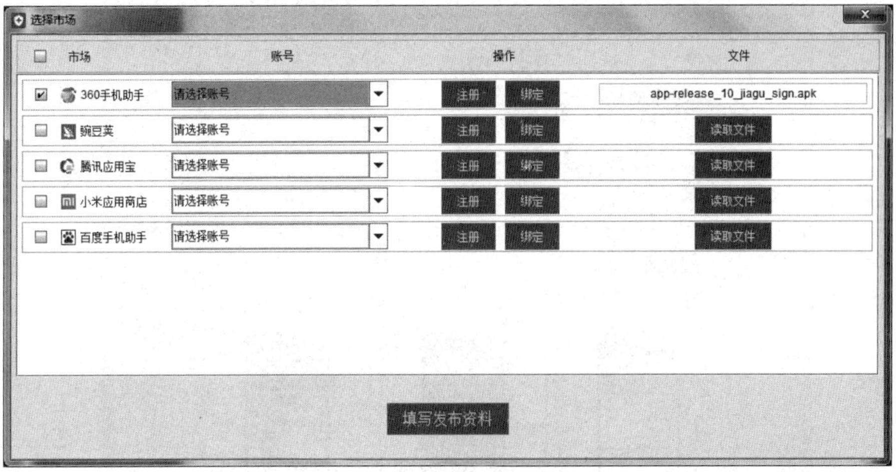

图 13-14　选择市场

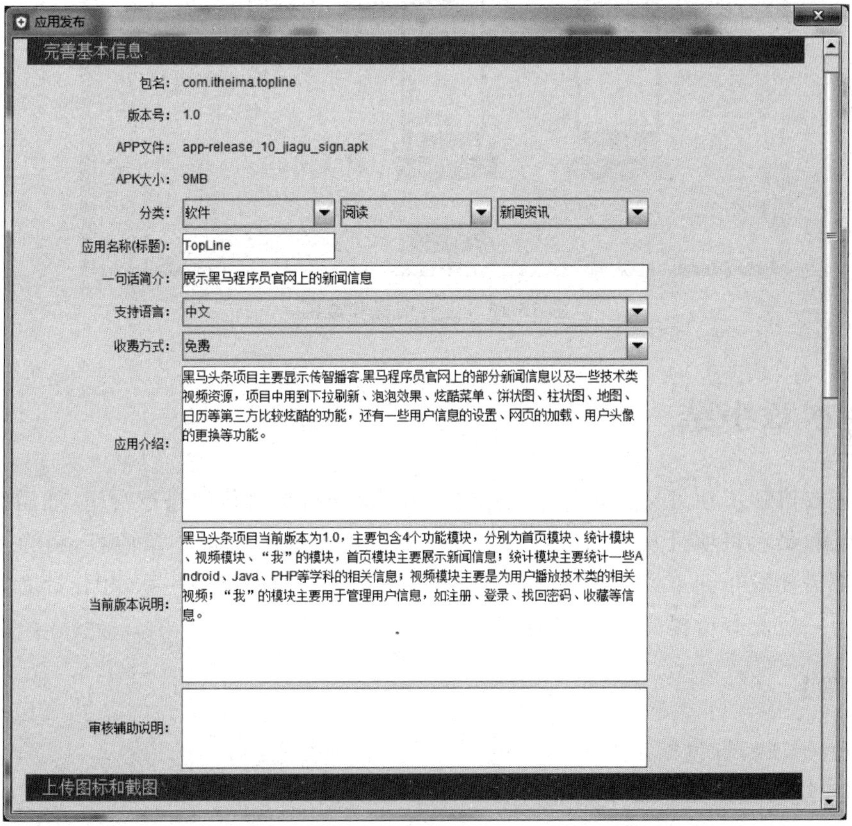

图 13-15　完善基本信息

在图 13-15 中，填写完成基本信息后，向下滚动可以看到"上传图标和截图"界面。在该界面中上传项目图标以及要展示的截图（图片必须按照要求上传），如图 13-16 所示。

项目信息填写完成后，单击"提交审核"按钮，项目进入审核阶段，当审核通过后，即可在市场中进行下载。

图 13-16　上传图标和截图

13.5　本章小结

本章主要讲解了项目从打包到上线的全过程，首先将项目代码进行混淆，然后将代码进行打包、加固，最后将项目发布到市场中使用。读者需要熟练掌握本章内容，因为企业中的每个项目都需要发布到市场中供别人使用。第 13 章是本书的最后一章，相信读者学完本书后在技术上一定能有所提高、收获颇丰。

【思考题】

1. 如何给项目打包？
2. 如何给项目加固并发布到市场？